The
Tao of Totality:
A Book About Space
& Time

By Paul H. Conner

DEDICATION

Dedicated to the memory of my parents.

CONTENTS

INTRODUCTION 1

    FATE 3
    PASSION.. 4
    PERSISTENCE 5

PART 1: PHILOSOPHY 33

    CHAPTER 1: ASIAN PHILOSOPHY 35
    CHAPTER 2: ASIAN ANTIQUITY.. 41
    CHAPTER 3: MODERN THINKERS 47

PART 2: AEROSPACE.. 51

    CHAPTER 4: ROCKET REALITY.. 53
    CHAPTER 5: REAL STARSHIPS 60
    CHAPTER 6: THRUST GRAVITY... 70
    CHAPTER 7: COMETS & ASTEROIDS 72
    CHAPTER 8: HACKING MARS... 76
    CHAPTER 9: SURFING & EXOPLANETS 82
    CHAPTER 10: NAKED WORMHOLES.. 92
    CHAPTER 11: HOLLYWOOD HYPERDRIVE 95
    CHAPTER 12: QUANTUM TELEPHONY 100

PART 3: COSMOLOGY 103

    CHAPTER 13: GRAND UNIFICATION 105
    CHAPTER 14: EVERYTHING THEORY.. 115
    CHAPTER 15: AFTERGLOW.. 130
    CHAPTER 16: THEORY DEFENSE 140
    CHAPTER 17: TOTALITY 144

APPENDIX.. 159

INTRODUCTION

FATE

With a heart of curiosity, not arrogance, I have pondered the "theory of everything." As a result, I have come to a conclusion. I don't have the high credentials, or celebrity status to make such a claim. As such, this book should be thought of as entertainment and perhaps comedy. However, I believe you will be shocked by the answer. Unless you already guessed that it's 42.

So, what are my credentials, beyond just being a curious dude? Well, I have a bachelor's degree from Florida Tech, 33 certificates from edX, and 2 aerospace engineering related U.S. patents. On the second patent, I was my own attorney throughout the whole process, so I have some legal skills too.

Obviously, none of those things qualify me to write this book. But, since I probably have the right answer, I've decided to write it anyway. Lucky you.

Brain Backup

This isn't my first stab at this. I previously wrote a first book manuscript, that as of the time I am writing this, is unpublished. The goal wasn't really to become an author, but to write down all my scientific knowledge in case something bad happened to my brain. Kind of like a backup file.

When I was young, my father had a significant stroke, and I have never been sure whether I inherited the genetic trait that caused it. I was tested, but the medical knowledge at that time was poor, in my opinion, so I've always had it in the back of my mind.

Well, I did apparently get brain damage, not from a stroke, but long-haul COVID-19. My original book manuscript is so complex, I would never be able to write it today. However, now that it's written, I can imagine how it could be better. I have also seen new places where my theory can go to make it more complete.

Because I have been neurologically changed by my health experience, it almost makes me like a collaborator with a fresh perspective. So, I self-collaborated on a new book.

Steppingstones

Gaining the knowledge to assemble a theory of everything is like a series of steppingstones across a stream. The stones end where human knowledge stops. The book I needed to complete my journey didn't exist, so I wrote it for myself. After reading my own book manuscript, I had an "aha moment." I figured out the final steppingstone to reach the other side! It may sound nuts, but it is what it is.

PASSION

I've always been a maniac about space. When I was in kindergarten, I was asked repeatedly, "What do you want to be when you grow up?" That same year I went to the theater and saw the original "Star Wars" for the first time. After that I knew what I wanted to do with my life. I wanted the job Luke Skywalker had. At five-years-old I had no idea that starships weren't real.

My parents tried to burst my bubble slowly. First they said, maybe in my lifetime there would be starships that could go from planet to planet and beyond. Later they said, maybe in my children's lifetime. Later still, they said perhaps in my grandchildren's lifetime. In adulthood I realized the answer was there are never going to be starships outside of fantasy, unless someone sets their mind to it properly. I had the idea that maybe that could be me.

My father was an engineer at the phone company. When I was young he told me how great Alexander Graham Bell was. One person had a revolutionary idea that changed life for everyone in the world. In doing it, he made himself a fortune, and also created thousands of jobs, like my father's. In my mind, to be an inventor was the way to change the world for the better.

The heart of the problem is propulsion technology. It's what stands between humankind and the greatest adventure imaginable. 21st century astronomers working in the fields of exoplanets and astrobiology have determined that there are likely many Earth-like planets in our galaxy, some likely nearby. In our own solar system there are three locations that may have some form of past or present life (Mars, Europa, and Enceladus). Every 20 years, NASA says in 20 years we will land an astronaut on Mars. Robotic probes and chemical rockets are not the path to a world where mankind can walk among the stars like giants.

After 25+ years of effort, and two patents, I have a solution. It's a mix of aerospace engineering and particle science. It probably won't go anywhere, but I had to try to be "that guy." You know, the person crazy enough to gamble everything for the chance of opening up mankind's best destiny.

I have collaborated with some of the best minds in the world, at NASA and elsewhere. Many people seem to be far more classically intelligent than me, but I have had an advantage that has allowed me to hold my own in the interactions. I have the one key ingredient. I can think way, way outside the box! Anyway, my passion is in the stars.

PERSISTENCE

Most people enjoy eating sausages, but as the saying goes, no one wants to see how they are made. In some cases, this is true of innovation. It can be an ugly and sometimes nonlinear process. It helps to be passionate about what you're doing, especially during the darker times. However, a little bit of irrational obsession helps things move along as well. When all hope is lost, obsessive persistence can help you push through to a solution.

Early Development

Upon realizing that my first career choice of being a Jedi starfighter pilot was not entirely realistic, I had the idea that I should become a scientist. So I told my father that I wanted to be a space scientist. He told me that perhaps I should consider becoming an astrophysicist. I was the only kid in the first grade that said he wanted to be an astrophysicist when he grew up. But here too there were issues. Astrophysicists, though they study space, they don't really get to go to space. In the days before Hubble, research really was mostly analyzing data, mostly through a ground telescope. So, I determined that perhaps the problem was really access to space. This would eventually send me on the path of the engineer.

TV

I didn't know any astrophysicists, so how was I going to take this childhood career choice for a test drive? The answer came through my TV. The second milestone in my obsession was when I was eight. Carl Sagan's TV series "Cosmos" had a strong impact on the way I view science. Watching these shows really captured my imagination. The biggest consequence was that I was exposed to the concept of cosmology, which is essentially studying the universe as a whole, as opposed to focusing on just specific aspects. So, I started to try to look at space like a cosmologist instead of like an astrophysicist.

This resulted in an interesting problem though. When I mentally looked at the universe as whole, I discovered that our understanding of the universe has a fundamental logic flaw! Okay, think of it like this. If you state that the universe is finite, then it occupies a limited amount of space. You say, well maybe it's infinite? But, if you go with the Big Bang theory, it requires that it's finite. If the universe is finite, then what is outside the universe? It's a subtle, perhaps even childish question, but the logical analysis is interesting.

If there is nothing outside the universe as science presents it to us today, then logically you can't say where the universe is! Thus, the universe must be nowhere. Fair enough, but something can't be located nowhere. Only nothing can be located nowhere. So, basically through proper logical deduction, and rational thought, I intellectually proved that if the model of the universe I was presented with was true, then the universe doesn't exist. As a consequence, that also means that I don't exist. What is the point of living if this is all just a weird dream?

So one way or another, the model of the universe that I was presented with was wrong. No one else I have ever met has been troubled by this. I intellectually started to drift away from science. No one knew the answers to the important questions, and it didn't look like I was going to be exploring the galaxy anytime soon.

Comedy

My brief career in comedy started in the fourth grade. Based on a Halloween costume choice, my teacher asked me to do a book report in my Steve Martin outfit. However, this would not be in front of the class, it would be on the school's new TV show. My father helped me put together a complete Steve Martin inspired comedy routine. My performance was well received, and I became the star of the TV show for two years. Over those two years I learned to be a professional level comedy performer. My career ended when I graduated elementary school. Later in life I would apply my skills in public speaking situations.

When I went to middle school, I was separated from most of the students that had seen the show, as they went to a different school. At this school they really didn't like comedians. I was strongly encouraged not to be funny anymore. As life went on things overall became increasingly unfunny. But, in my adult life "Steve Martin" pops out every once in a while, in striking contrast to my otherwise reserved demeanor.

Science Fair

In time I matured in more practical terms. However, when I started high school I started getting interested in science again. It was getting closer to the 21st century and all the great promises of the world of tomorrow looked like they might not come true. I thought what is the point of living to see "the future" if there are no flying cars and space colonies, and all the other silly things promised by the popular sci-fi movies of the time. So, I thought well maybe I will pick something, only as a thought experiment, and see if I can figure it out. The simplest, at the time, seemed to be the flying car.

I was wrong. I thought that people were lazy, and that if you tried hard enough the problem would just work itself out. I played with the idea in my off hours for about a year. I even tried to recruit others to work on my project, but they were smart enough not to get involved. It ultimately was a total flop, and on the one year anniversary of this bad idea, I decided to quit. I thought I should redirect my energy back into my music aspirations, and try to increase my social standing at school. Then something crazy happened.

It came in a vision. It's easiest to say it was like a religious experience. I was watching a movie at home one evening. At the conclusion of the movie, the credits were playing and the room was very dark. I thought to myself, this is the moment when I am going to give up childish things and really focus on my future. This idea of a flying car that I had pursued over the course of exactly one year was pointless. It had clearly degraded into nonsense. I had been contemplating the idea of somehow propelling the car with electricity. At that moment I had the strangest epiphany.

In my mind I saw the flying car engine transform into a spacecraft engine. Now I was standing in outer space, above the Earth. I saw a large spacecraft slowly breaking planetary orbit. It looked as if it was lighting up. I intuitively knew that the engine that I had been just contemplating was powering this craft. There was a pregnant pause and the spacecraft took off so fast that I couldn't see it any more. What I was left with was the feeling of exhilaration of having witnessed this vision. I came back to the room I was in with a new life mission. It was to figure out how to make that spacecraft engine.

It took many years to finally figure out the secrets so efficiently contained in that vision. What kept me going was not the idea of wealth and glory, nor enhanced self-worth, but it was the fact that every time I seriously pursued this development, I would get the feeling I had in the vision. Ultimately, I was chasing this feeling more than the idea of the spacecraft engine itself. It had become a spiritual quest.

At this point in my life, I was a sophomore in high school, and my main problem in life was the fact that my science teacher was making me do a mandatory science project. Perhaps ironically, I had no interest in doing such a project and I was looking for a way to get through it as painlessly as possible.

I had the genius thought that maybe I could figure a way to take my new spacecraft propulsion concept and configure it for use as my science project. Ultimately it seemed like less work than the suggested topics I was given by my teacher. All I had to do was to convince her into letting me do this as my project.

I met with her after school in her office. I presented my idea, and she went for it. I had an easy out for my science project. Then she said something strange. I have to paraphrase, because it has been some time ago. Basically, she said that I had just pitched a patent not a science project. I didn't take her seriously. I thought to myself, yeah right. I couldn't imagine ever getting a patent.

So now that my mystical vision is now officially a high school science class project it needed a name. After a lot of stupid ideas, I decided to just keep it simple: The Electron Propulsion Engine. Science project presentation day came, and I presented my idea. My comedy persona made a rare appearance, and it was received by all very favorably. Mission accomplished, job well done, time to move on. But, my science teacher said my project was so good that I should put it in the school's science fair. I think I got peer pressured into it. The work was done; all I had to do was show up at the fair, so I said yes.

NASA

The high school science fair experience was okay. The school I was at was actually into science. The high school I was originally supposed to attend, before my family moved over the summer before my freshmen year, had only three projects for the entire school. We're talking about a school with a population greater than 2,000 students. If I had entered my project in the fair there, I would have taken first prize, and then been tarred and feathered. But at my current high school, science was actually socially acceptable.

Things didn't go great however, I only got honorable mention. The judges found my project a little hard to mentally get hold of. In a quirk of fate, I was asked to advance to the county science fair, because they didn't have enough participants in the engineering division. So, now I am going to the county fair.

Generally speaking, the county experience didn't go well. The judges didn't get my project, and ultimately I didn't place in the county fair. However, there were these two guys that were really friendly, and came and asked me a lot of questions. They seemed to know a lot more than the other judges. When I didn't place in the fair, I assumed that they were really just interested spectators.

Before the award ceremony, the projects were marked as to how they placed. In front of mine was nothing. I was ready to go. My father said he had a strange feeling, and that we should go to the award ceremony. When to ribbons where given out: first, second, third, and honorable mention, as I thought, I didn't get anything. Then they gave out special awards. The presenter representing NASA walked up to the podium. He called out the names of the winners of the NASA awards, and asked them the come to the front of the full auditorium. I was given the NASA first prize award for outstanding achievement. This was a turning point in my life. This definitely helped me choose the direction of my career choice.

University

I chose a college that was close to NASA. Somehow I thought that engine exhaust equaled excellence. At the time I believe there was no university located closer to the Kennedy Space Center than the Florida Institute of Technology. In fact, it had actually been started by NASA as an engineering school in the '50s before it had been converted to a private college. I showed up wide-eyed at Florida Tech after receiving my acceptance to the School of Aerospace.

My college years had mixed results. My grades were poor, and my future as a rocket scientist was in question. Ultimately in order to stay at Florida Tech I would have to change my major. However, I had a final act of brilliance.

In my sophomore year at university, I knew I was going to have to give up on the dream of being an aerospace engineer. But, as one door closes, another door opens. I decided that maybe there was another way. Perhaps I could become an aerospace renegade instead.

One day at the end of the semester, I waited outside of my physics professor's office. Over the past two years, although my grades were not good enough to stay in the program, I had learned a lot. The level of refinement and miniaturization of the Electron Propulsion Engine had reached maturation. As my last act as an aerospace student, I typed up my invention. When it was my turn to talk to the professor, I handed him my project. He looked at it quietly, and then he turned a little pale. He reached on a shelf for a book, and started reading very fast, then looked at my paper again. Finally, he looked up and said that he couldn't think of a reason why it wouldn't work. He asked how I could be failing his class? We talked a little longer and he ultimately agreed with my decision to change majors, but he insisted that I go buy the more advanced physics book that corresponded with the more advanced level classes. He put forward the theory that perhaps I had a different learning style than the other students and that I should continue learning physics on my own.

Encouraged in the validity of my invention, the same day I visited one of my other professors. This time it was my mechanical engineering professor. My theory in seeking the advice of my professors was that I figured if I screwed something up, they would be more than happy to point it out. He took it and looked at it for a long time, then slid it across the desk and said, "good luck!"

Next stop the post office. I had already had a large envelope with me in my backpack. I ran down the stairs of the science tower where the professors' offices were. It was a brief hike from the tower to the post office and I got intercepted by one of my other professors whom I owed an assignment, which I also had in my backpack. It was almost closing time and I made it by seconds.

The Electron Propulsion Engine brief made it back home, where my mother had a notary notarize it. A couple weeks later, I returned home for Christmas break, and the search for a patent attorney began.

When I went back to Florida, a patent attorney had been hired, and I was now a technical and scientific communication student. The plan was that I would use my skills after graduation to promote and sell the invention. I thought this new major would give me the skills I would need to do just that. My aerospace curriculum had been very demanding, but now that I was a humanities student there was a little more latitude. This allowed me to have different opportunities like taking Japanese.

The patent process resulted in some dark days. I had a very good patent attorney. He changed the name of the invention to the Electron Propulsion Unit, in the hope that it would make the patent more marketable. Just about every bad thing that can befall an independent inventor occurred. It was a terrible time for me because it took years to work out and it was at the most critical point in my college education. By the time I was granted the patent, I felt like my own co-counsel. I feel like I graduated with two degrees, with one being in patent law.

Patent

To make a long and uninteresting story short, I will just say things could have gone better. My patented ideas were met with open hostility and my skills were very much not in demand in the job marketplace. However, I learned the concept of "iteration."

With every spectacular failure, I improved iteratively, both my product and myself. Over time I started to get really good at my craft. My invention was presented in a way that was so tight, no intelligent argument could be made against it. When the rejections were impossible, those in industry simply stopped responding. The only safe answer for others was no. This definitely was the time for me to tactically retreat.

While I was regrouping, I decided to hire myself. I finally took the plunge and formally founded my own aerospace company, Conner Creations, LLC. Since I was the only employee, I got to be CEO. That is unless of course you count my dog. Dachshunds can be very helpful to the creative process.

Business cards aside, this was not much of a strategy. At some point I just said to myself, I am just not smart enough to solve my problems. So, I devised a way to get smarter for free.

Reeducation

I pretty much used the Internet to rebuild my brain. Initially I used Apple's iTunes U to go back to college. The only difference was that I got to coordinate my own education. I thought, what makes me curious, what skills do I lack? The eclectic nature of my own curriculum could never practically be duplicated if I enrolled again at a physical university. At first, I audited everything, so no grades, no homework, just results. I went fast and just saw what would stick in my mind, and what just wouldn't go in.

But it wasn't just a hectic and eclectic schedule that did the trick. I got to choose from the best universities and teachers on Earth. That's education. Here is just a sampling of what I took. Below is a downselect from the total of 42 courses.

Sample of Audited iTunes U Courses (Alphabetically):

1. Approaching Shakespeare, Oxford
2. Astrobiology & Space Exploration, Stanford
3. Astrophysics, Yale
4. Basics of Culinary, The Art Institutes
5. Classical Mechanics, MIT
6. Electricity & Magnetism, MIT
7. Exploring the Evolution of Language, TED (Compilation)
8. Exploring the Geometry of Form, TED (Compilation)
9. Game Theory, Yale
10. General Philosophy, Oxford
11. Justice with Michael Sandel, Harvard
12. Leading Wisely, TED (Compilation)
13. Mastering Tech-Artistry, TED (Compilation)
14. Quantum Mechanics, Oxford
15. Quite Easily Done, Cambridge
16. Science, CERN
17. Solid State Chemistry, MIT

18. Stargazing, Oxford

19. The Dalai Lama at Stanford, Stanford

20. The Edge of Knowledge, TED (Compilation)

21. This is CS50, Harvard

In addition, I read a lot. I read the "King James Bible" cover-to-cover. I also read other Judeo-Christian texts like "The Nag Hammadi Library" and "The Book of Enoch the Profit."

I read biographies such as "Einstein: His Life and Universe" and "Steve Jobs." It gave me more perspective. I guess all innovators are treated like crap at some point. So I learned to take it less personally.

I read some of the science books of Michio Kaku: "Physics of the Impossible" and "Physics of the Future." I wanted to guarantee that my perspective on the science of the future wasn't in any way dated.

Having completed the curriculum, I set out for myself, I thought, okay now I'm ready. Gaps filled; knowledge updated. But then something very strange happened, unity!

Unity

When I was in my early twenties, I thought it would be cool to one day come up with the grand unification theory, what people now days generally call the theory of everything. It seemed like an impossible goal, but it was fun to think about. My father and I would talk for hours about the nature of the universe on our family deck late at night.

Far more recently, I was watching a Yale astrophysics class on the iTunes U App on my iPad. The professor was drawing a graph of the movement of the universe. Then he stopped, and I said to myself, "no, keep going, it's not finished." In that moment, I realized I could put together a complete theory of everything!

The mostly negative experience I had trying to sell the Electron Propulsion Unit had yielded tons of technical improvements to the propulsion system. Additionally, watching an iTunes U lecture from Stanford, I got the idea for a completely new application for my technology. Collectively, this set me up for a new patent. This time I would go back to the original name, the Electron Propulsion Engine.

Now back to the theory of everything. While I was working out the details, I realized the device I needed to test my completed theory of everything was in fact the Electron Propulsion Engine. It's full circle. I achieved cosmic unity and personal unity, as all the random pieces came together to make one thing.

I applied for a new patent. I wrote a scientific paper recording the theory of everything. Now all that was left was to write down how it was all done. Then came my final and ultimate challenge: writing a book!

Synchronistic Event

I sent a copy of the original version of my theory of everything to the MIT professor of the iTunes U course I had been auditing. In my email, I apologized for bothering him in his retirement. He responded that he had come out of retirement and was now teaching courses at something called edX.

Edx is an online MOOC platform, founded by Harvard and MIT. A MOOC is of course a massive open online course. It provides courses from the world's premiere universities. Given that the institution is 100% online, anyone can attend. Entrance exams and competition are based on the model of limited physical space.

Innovation happens when disciplines are mixed, so a linear path to a degree stifles and even can prevent innovation. At edX, I have taken a wide variety of courses from the music business to astrophysics. No one else may see the common thread in my choices, but I see it! Everyone who wants to be an innovator should have this same opportunity in the 21st century!

edX Adventure

Oh, the places I went! I studied astrophysics in Australia. I studied chemistry in Kyoto. I studied history at Harvard. I studied Chinese in China. Given that this was a major part of my preparation to write the final iteration of this book, I have listed below my curriculum.

Certificates of Mastery (Chronologically):

1. HarvardX: China (Part 1): The Political & Intellectual Foundations of China

2. TsinghuaX: History of Chinese Architecture: Part 1

3. HarvardX: China (Part 2): The Creation & End of a Centralized Empire

4. HarvardX: China (Part 3): Cosmopolitan Tang: Aristocratic Culture

5. BerkleeX: Introduction to the Music Business

6. HarvardX: China (Part 4): A New National Culture

7. MITx: Entrepreneurship 101: Who is your customer?

8. CornellX: Wiretaps to Big Data: Privacy & Surveillance in the Age of Interconnection

9. HarvardX: China (Part 5): From Global Empire to Global Economy

10. TsinghuaX: Historical Relic Treasures & Cultural China: Part 1

11. ANUx: Greatest Unsolved Mysteries of the Universe

12. TsinghuaX: History of Chinese Architecture: Part 2

13. HarvardX: China (Part 6): The Manchus & the Qing

14. KyotoUx: The Chemistry of Life

15. HarvardX: China (Part 7): Invasions, Rebellions, & the End of Imperial China

16. HarvardX/MITx: Visualizing Japan (1850s-1930s): Westernization, Protest, and Modernity

17. HarvardX: China (Part 8): Creating Modern China: The Birth of a Nation

18. BerkeleyX: The Science of Happiness

19. HarvardX: China (Part 9): Communist Liberations

20. BUx: Alien Worlds: The Science of Exoplanet Discovery & Characterization

21. HarvardX: China (Part 10): Greater China Today: People's Republic, Taiwan, & Hong Kong

22. HarvardX: Super-Earths & Life

23. UC3Mx: Explaining European Paintings, 1400 to 1800

24. MITx: Entrepreneurship 102: What can you do for your customer?

25. BerkeleyX: Electronic Interfaces

26. MITx: Introduction to Aerospace Engineering: Astronautics & Human Spaceflight

27. TenarisUniversityX: Introduction to Steel

28. AdelaideX: The World of Wine

29. HKUx: Humanity & Nature in Chinese Thought

30. MandarinX: Chinese Language: First Steps in Chinese

31. MandarinX: Chinese Language: Learn Basic Mandarin

32. BerkleeX: Introduction to Music Theory

33. HarvardX: CS50: Introduction to Computer Science

There were a number of highlights to my academic adventure of intellectual reinvention. I will share a few.

Astrophysics

As an elementary school student, I had dreamed of studying to be an astrophysicist. I never did due to my later interest in aerospace engineering. Through edX I was able to not only study astrophysics, but learn from the Nobel laureate who discovered dark energy, Prof. Brian Schmit. Although I found parts of the course difficult, it was very gratifying when I got a perfect score on the final exam.

Astronautics

Although I did study aerospace engineering at my alma mater, I thought it would be interesting to test my somewhat older brain. This time, however, I got to take it at MIT. Also, I got to learn from astronaut, Prof. Jeffery Hoffman. Oh, and unlike the first time around, I got a 100% on the whole course. That was confidence building.

Exoplanets

When I was at university, if you were interested in planets around other stars, that meant you liked science fiction, because there was no scientific evidence to support that possibility. However, today with the discovery of many planets around other stars (exoplanets), it's a legitimate scientific pursuit. I was very lucky to have the opportunity to take a couple of exoplanet courses.

I took a course called "Alien Worlds." It was one of the most fun standalone courses I have taken. Partially because I like space stuff a lot, and partially because the course had live online office hours. I asked the craziest things I could think of just to screw with the professor's mind. Some of the best ones are paraphrased as follows:

1. For example, I asked stuff like if you were sunbathing at the beach on a Goldilocks Super-Earth around an M dwarf star, would you need to take sunscreen?

2. Given that M dwarf stars have a predicted lifespan that exceeds the lifespan of the entire universe, are they perpetual motion machines, which violate the first and second laws of thermodynamics?

3. Can high-gravitational objects in the galaxy disturb the Oort cloud and send a swarm of killer comets heading towards the Earth? If so, how long would it take from the initial disturbance for the Earth to blow up?

4. If you put a space telescope at a distance of 550 AU (1 AU = the distance between the Earth and Sun), you could use the Sun as a gravitational lens based on Einstein's theory of general relativity. Given that this would be equivalent to a telescope with a mirror diameter of one million miles, would the resulting resolution allow for the direct viewing of aliens in their natural habitat?

According to the professor, my third question was an actual graduate course final exam question! I'm sure he worded it a little different though.

Pharmaceuticals

One of the unique opportunities I had was to take a course from Japan. In this Kyoto based biochemistry course, I was able to be creative in a new discipline. I think the motivation behind this course was actually to crowdsource ideas for future pharmaceutical research. As such the assignments were to submit such ideas after being given the necessary intellectual tools to do so. My first idea was as follows:

Arteriosclerosis/Atherosclerosis Familial Hypercholesterolemia Aptamer:

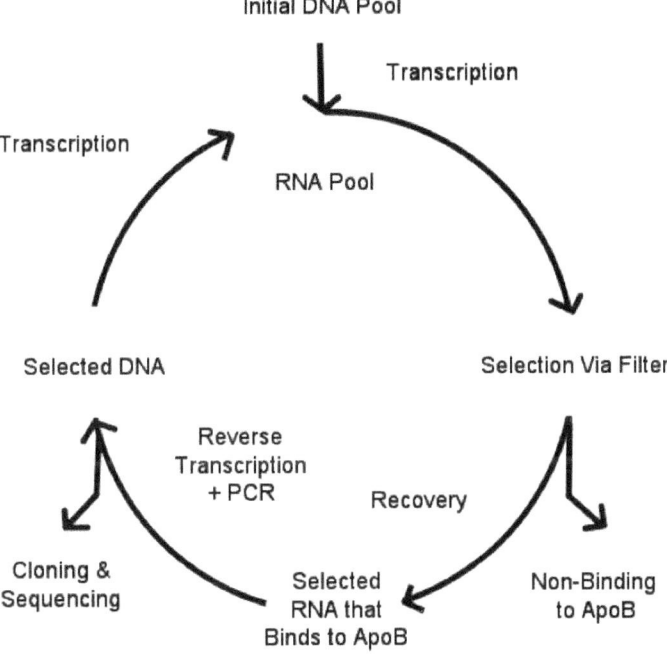

My idea adapts SELEX technique for curing macular degeneration, by applying it to a combined cure for both arteriosclerosis and atherosclerosis, where a family-passed mono-gene is responsible for hyper-cholesterol production. This will be done by substituting anti-VEGF RNA aptamer for anti-apolipoprotein B (ApoB) RNA aptamer, the counter to the gene responsible for the disease causing familial hypercholesterolemia (FH). Turning off this mutant gene would benefit 15 million people. The cure will be administered directly into the blood stream by IV therapy or injecting it directly into the liver. I chose this because FH arteriosclerosis killed my father at 52.

I had spent many hours of my youth speculating on a possible cure for this disease, and formulating a realistic strategy gave me a slight sense of closure. My second idea which ironically received high praise from my peers was a little less serious:

Get-Home-Safe Pills: Synthesized Proteins Breakdown Alcohol in Blood:

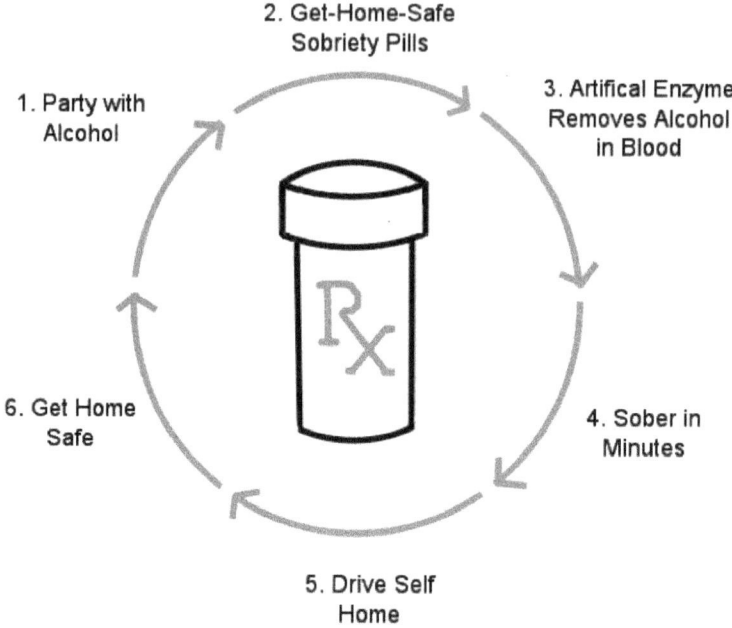

Attending events where alcohol is served involves prosocial behaviors and medicinal stress relief. However, simply returning home has severe social and legal liabilities. Get-Home-Safe instant sobriety pills utilize parallel synthesis to create artificial enzymes that breakdowns alcohol in blood more efficiently than naturally possible. Liver enzymes naturally remove alcohol at a fixed rate of 0.016 BAC/hour. If this drug could remove alcohol at a magnified rate of 0.16 BAC/hour, inebriation could be neutralized in minutes. It also has the combined benefit of preventing alcoholic liver disease. It's the pill for after the party, so that you can get home safe.

Robotics

I tried my hand at robotics through a UC Berkley course. Along with thousands of other participants, I got to follow along with the two co-professors of the course, and make what came to be known as "Bouncy Bot." Using the parts supplied in a kit, I made a robot that responded to light, sound, made cool R2-D2 noises, and bounced up and down.

Berkeley Bouncy Bot:

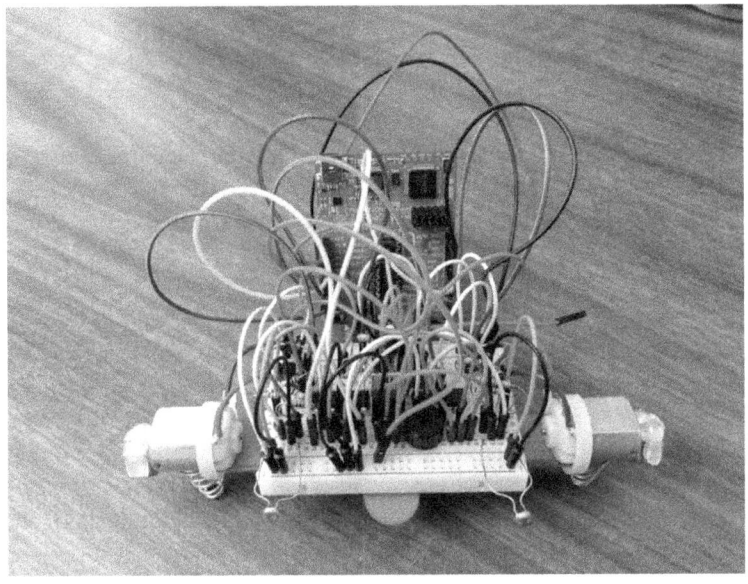

I don't think I have a career in robotics, but it worked and I got a perfect score on my project. The construction style definitely had that mad scientist je ne sais quoi.

Harvard?

If I could go back in time and tell the teenage version of myself that one day I would study at Harvard, I would have thought I was crazy. Back then I found academics very difficult. The only way I would have seen a future path to Harvard would have been to win the lottery, and then to get a brain transplant. I would have never imagined that broadband Internet would allow me to have this opportunity for free, and that I would be really good at it. Not to sound braggadocios, but when I quit edX, I had a straight A average at both Harvard and MIT. I guess I was a late bloomer.

My two favorite courses at edX where both given by Harvard. I can't really choose a favorite, because they were so different from each other, and so unique in their own way.

ChinaX

This course was truly revolutionary. Firstly, there was almost 50,000 students actively participating in this online course. Secondly, this was the longest MOOC in history! It was a ten-part course of study that spanned 18 months, consisting of 51 modules, and requiring 5-6 hours of study per module. There were student participants from 167 countries, with 70% of the students from outside the U.S., and 15% from China.[1] Like many students, I stuck with it until the end because I was absolutely hooked.

The course was about the history of China. I mean all of it. That's from the Neolithic period to the present day. Technically the course even discussed the future all the way to 2035!

A feature that made this course unique was the "office hours." The professors teaching the course held a video session after each module. In it they would respond to the discussion question responses submitted through the discussion forum. This feedback made the course dynamic and interactive.

The challenge of writing interesting responses that would get mentioned in this Ivy League environment, resulted in an improvement of my writing skills. My pithy quips are now better than ever. I managed to get mentioned in 10% of the 50 office hours. This included the honor of being chosen as the final comment for the entire ChinaX series. As the Chinese would say, I received a crazy amount of "face" for that. Even though it was done anonymously, through my nom de guerre PaulHC, that was one of the top academic moments of my life. It's not often that you get singled out in a classroom of 50,000!

This is CS50.

Clearly this is the coolest course at Harvard, both in person or online! Where do I start? The official name of the course is: "Introduction to the Intellectual Enterprises of Computer Science and the Art of Programming." However, it's world-renowned as simply, CS50. As of the end of 2015, this course had taught 800,000 edX students to code.[2] About 72% of students both on and off campus have no prior experience programming.[3]

The lectures are nothing short of theater. It's the only course I've ever seen with an in-house EDM DJ! The whole thing is just wild. CS50 was featured prominently in the CNN documentary film "Ivory Tower," as a positive example of what the future of education could be. This is the only class I've taken that has had its own catchphrase: "This is CS50."

The course was composed of 9 problem sets and 1 final project. The problem sets included stuff like:

- Building Mario's pyramid, from the Nintendo Super Mario Bros. videogame.
- Learning cryptography by encrypting and decrypting messages.
- Implementing the classic Atari videogame Breakout.
- Learning forensics by recovering deleted photos.

- Taking a dictionary of 140,000 words, and creating the fastest spellcheck possible.
- Implementing a web server, that serves up actual web pages.
- Creating a web application, that can buy and sell stocks with virtual dollars, programmatically using Yahoo finance to retrieve nearly real-time prices.
- A mashup of Google Maps and Google News, to create a web application, with which to find web articles from nearly any city in the world.

The course used a combination of several programming languages including: Python, C, JavaScript, SQL, CSS, and HTML.

There were two great highlight moments for me. The first was the opportunity to step into the shoes of Apple co-creators Steve Jobs and Steven Wozniak. The problem set was to recreate the classic Atari 2600 videogame "Breakout," which they made.[3] However, instead of physically wiring a circuit board, the problem set was done completely in software with the C programming language.

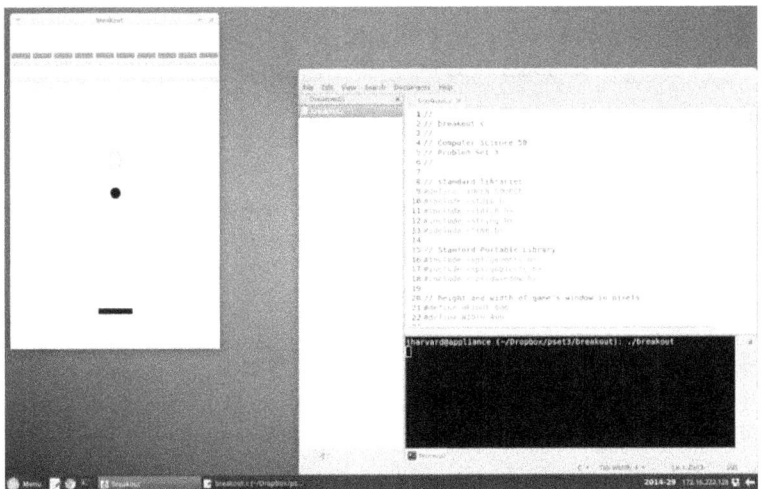

When I did this problem set, CS50 was using something dubbed "the appliance." This was a hypervisor that allowed for different computer systems to have the same virtual machine Linux OS. At present, CS50 has moved to the next level experience. Now through CS50 IDE, students can code directly into "the cloud."

The second highlight was my final project and corresponding video presentation. This resulted in my final career invention: Instant Spy™ software. This was the perfect conclusion to my academic adventure.

Unbreakable Code

Once upon a time, back in the year 2012, a British couple were renovating their chimney. Inside they found the skeleton of a WWII carrier pigeon, with a secret coded message tied to its leg. It was sent to GCHQ, the code breaking arm of MI5 and MI6. There is no method to break the code.[4] The code's key is the length of the message, with the message being 135 letters long (shifted randomly between 0-25). The total keys: $26^{135} = 1.05 \times 10^{191}$! The total number of combinations is about a googol squared. It would take today's computers thousands of years to check all possible combinations. Additionally, decryption generates every possible message of 135 letters long, with all being equally likely. Thus, it's 100% impossible to break.[5]

WWII Code:

AOAKN HVPKD FNFJW YIDDC
RQXSR DJHFP GOVFN MIAPX
PABUZ WYYNP CMPNW HJRZH
NLXKG MEMKK ONOIB AKEEQ
WAOTA RBQRH DJOFM TPZEH
LKXGH RGGHT JRZCQ FNKTQ
KLDTS FQIRW AOAKN 27 1525/6

Quantum Computer

In January of 2014, CNN confirmed that the NSA was building a quantum supercomputer through a subcontractor.[6] This system is designed for nothing other than cracking all known online encryption.

In a classical computer one bit equals one piece of information, a 0 or a 1. A quantum bit, referred to as a qubit (pronounced "cubit"), can be a 0 or a 1 or a "superposition" of the two. The result is that quantum computers can be exponentially more powerful than a classical computer. For example, 300 classical computer bits equals 300 pieces of information, such that the amount of data equals "n." For 300 qubit particles the amount of information equals 2^n. In other words, for 300 particles the total information equals 2^{300}, which is equal to the total number of all particles in the universe! The trick is to have the qubits in the "fully entangled state."[7]

n Qubits = 2^n Classical Cumputer Bits

A photon, a nucleus, or an electron can be used as a qubit. To minimize the energy of the system, liquid helium is used to lower the systems temperature within a few hundredths of a degree from absolute zero. The spin is aligned with a magnetic field from a superconducting magnet. Spin down is low energy ground state (0) and spin up takes energy (1). Information is written by flipping the qubit spin up with a pulse of microwaves. A specific frequency is used relative to the magnetic field the particle is sitting in. The frequency in which an electron qubit responds to depends on the spin direction of its nucleus. The nucleus can be made to act as a frequency selector. The qubit can be in both spin states at once, or in quantum superposition, but only before it's measured. The process reduces the number of steps required to complete a computation. To work, the qubits need a home with no nuclear spin that's completely nonmagnetic, in other words a pure crystal of silicon-28.[8]

Quantum computers are not a replacement for classical computers. They're not universally faster, only faster for special types of calculations. If you're watching YouTube in HD or writing a document in Word, a quantum computer will not give you any improvement if you're going to use classical algorithms to get a result. These operations are not going to be faster; in fact, they're probably going to be slower.[7]

However, for specific operations the quantum computer is exponentially faster. For example, to find the prime factors of a 2048-bit number it would take a classical computer millions of years. However, a quantum computer could do it in just minutes! This is because qubits take advantage of quantum superposition to reduce the number of steps required to complete the computation. Thus, most of today's encryption could be cracked essentially instantaneously.[8]

Quantum Adversary

The most widely used algorithm for encrypting data on the internet is called RSA. Basically, it's the factoring of very large integers. Security is based on how long it would take for an attacker to crack the message. An attacker using a classical computer would take an unrealistically long time to factor large numbers. RSA typically uses a 1024-bit number. RSA would be broken if a fast way to factor integers was found. In theory if they had access to a faster method, then someone could read all data encrypted with RSA.[9] Quantum systems are exponentially powerful.[10] One thing that quantum computers do really well is break cryptosystems. Quantum computers can break much of modern cryptography, especially RSA.[11] You may say well that's okay as long as only the "good guys" have this technology. However, if history is any guide, soon adversaries and competitors of the U.S. will have this technology as well. This is a serious problem in the making.

Digital Solution

My project solves the problem of a world without viable encryption by digitizing the one-time pad cipher WWII technique. It's a natural solution given that a tweet on Twitter use to be 140 characters or less. An unbreakable code that's 135 letters long plus a numerical code key number is almost too perfect to be true. It proves that, sometimes the old ways are the best.

Instant Spy™

The complication is that the initial key can't be created by a computer, because it has to be truly random. So, this process has to remain partially analog.

Generate, Mutate, & Burn

The solution I came up with is the "Double-O Encryption Process" which is based on three concepts: generate, mutate, and burn. The first step is to create the initial code key. It has to be truly random, so let's mentally go to Las Vegas and use some playing cards.

Professional Poker Shuffle:[12]

1. Wash Shuffle for 7 Seconds
2. Riffle Shuffle Twice
3. Box Shuffle
4. Riffle Shuffle Again
5. Cut the Cards

After drawing 15 cards, recording them, and reshuffling 8 times, I performed the whole procedure in 33 minutes and 46 seconds.

Input / Output

Ace of Spades = A	6 of Hearts = S	Jack of Diamonds = K
2 of Spades = B	7 of Hearts = T	Queen of Diamonds = L
3 of Spades = C	8 of Hearts = U	King of Diamonds = M
4 of Spades = D	9 of Hearts = V	Ace of Clubs = N
5 of Spades = E	10 of Hearts = W	2 of Clubs = O
6 of Spades = F	Jack of Hearts = X	3 of Clubs = P
7 of Spades = G	Queen of Hearts = Y	4 of Clubs = Q
8 of Spades = H	King of Hearts = Z	5 of Clubs = R
9 of Spades = I	Ace of Diamonds =A	6 of Clubs = S
10 of Spades = J	2 of Diamonds = B	7 of Clubs = T
Jack of Spades = K	3 of Diamonds = C	8 of Clubs = U
Queen of Spades = L	4 of Diamonds = D	9 of Clubs = V
King of Spades = M	5 of Diamonds = E	10 of Clubs = W
Ace of Hearts = N	6 of Diamonds = F	Jack of Clubs = X
2 of Hearts = O	7 of Diamonds = G	Queen of Clubs = Y
3 of Hearts = P	8 of Diamonds = H	King of Clubs = Z
4 of Hearts = Q	9 of Diamonds = I	
5 of Hearts= R	10 of Diamonds = J	Joker = Ø

Standard 52-Card Deck: Input of Rank & Suit = Output of Alphabet Value

As cool as playing with cards is, this process can be tedious. The magic of my concept is in the mutation. For the effort of creating 1 truly random code key, a codebook of 100 is created! The secret is the Instant Spy Algorithm™.

Instant Spy Algorithm™

I Invented the following on February 15, 2014. It's a byproduct of a lot of trial and error. The truly random key is generated by the user's input = Spawn. Spawn is divided into 5 segments designated A, B, C, D, and E. From the original user generated key, 10 keys are created by reordering the 5 segments. In Stage #2, 60 total keys are created from the 10 keys of Stage #1. This is done through a sequence of keys encrypting keys and mirroring some of the preceding encrypting. Stage #3 divides again the original key and scrambles it, allowing 40 total additional keys to be created in Stage #4, in the manner of Stage #2. The result is 100 truly random keys from only 1 key input sequence by the user. The possibility of a pattern being detected in the mutant one-time pad cipher keys is further negated by the pseudorandom selection of these keys for encryption, such that not only are they only used once, but in non-sequential order, obscuring the method used.

Mutation Stage #1:

Spawn =	A, B, C, D, E
Leonardo =	B, C, D, E, A
Raphael =	C, D, E, A, B
Donatello =	D, E, A, B, C
Michelangelo =	E, A, B, C, D
Mirror1 =	E, D, C, B, A
Mirror2 =	A, E, D, C, B
Mirror3 =	B, A, E, D, C
Mirror4 =	C, B, A, E, D
Mirror5 =	D, C, B, A, E

Mutation Stage #2:

Alpha1 = Spawn ← Spawn
Alpha2 = Leonardo ← Spawn
Alpha3 = Raphael ← Spawn
Alpha4 = Donatello ← Spawn
Alpha5 = Michelangelo ← Spawn
Alpha6 = Mirror2 ← Spawn
Alpha7 = Mirror3 ← Spawn
Alpha8 = Mirror4 ← Spawn
Alpha9 = Mirror5 ← Spawn
Alpha10 = Mirror of Alpha1
Alpha11 = Mirror of Alpha2
Alpha12 = Mirror of Alpha3
Alpha13 = Mirror of Alpha4
Alpha14 = Mirror of Alpha5
Alpha15 = Mirror of Alpha6
Alpha16 = Mirror of Alpha7
Alpha17 = Mirror of Alpha8
Alpha18 = Mirror of Alpha9

Bravo1 = Leonardo ← Leonardo
Bravo2 = Raphael ← Leonardo
Bravo3 = Donatello ← Leonardo
Bravo4 = Michelangelo ← Leonardo
Bravo5 = Mirror3 ← Leonardo
Bravo6 = Mirror4 ← Leonardo
Bravo7 = Mirror5 ← Leonardo

Bravo8 = Mirror of Bravo1
Bravo9 = Mirror of Bravo2
Bravo10 = Mirror of Bravo3
Bravo11 = Mirror of Bravo4
Bravo12 = Mirror of Bravo5
Bravo13 = Mirror of Bravo6
Bravo14 = Mirror of Bravo7

Charlie1 = Raphael	←	Raphael
Charlie2 = Donatello	←	Raphael
Charlie3 = Michelangelo	←	Raphael
Charlie4 = Mirror4	←	Raphael
Charlie5 = Mirror5	←	Raphael

Charlie6 = Mirror of Charlie1
Charlie7 = Mirror of Charlie2
Charlie8 = Mirror of Charlie3
Charlie9 = Mirror of Charlie4
Charlie10 = Mirror of Charlie5

Delta1 = Donatello	←	Donatello
Delta2 = Michelangelo	←	Donatello
Delta3 = Mirror5	←	Donatello

Delta4 = Mirror of Delta1
Delta5 = Mirror of Delta2
Delta6 = Mirror of Delta3

Echo1 = Michelangelo	←	Michelangelo

Echo2 = Mirror of Echo1

Mutation Stage #3:

Splinter1 =	A, C, B, E, D, E, F, G, H, I, J, K, L, O, N
Splinter2 =	N, A, C, B, E, D, G, F, I, H, K, J, M, L, O
Splinter3 =	O, N, A, C, B, E, D, G, F, I, H, K, J, M, L
Splinter4 =	L, O, N, A, C, B, E, D, G, F, I, H, K, J, M
NewMirror1 =	N, O, L, M, J, K, H, I, F, G, D, E, B, C, A
NewMirror2 =	O, L, M, J, K, H, I, F, G, D, E, B, C, A, N
NewMirror3 =	L, M, J, K, H, I, F, G, D, E, B, C, A, N, O
NewMirror4 =	M, J, K, H, I, F, G, D, E, B, C, A, N, O, L

Mutation Stage #4:

Foxtrot1 = Splinter1   ←   Splinter1
Foxtrot2 = Splinter2   ←   Splinter1
Foxtrot3 = Splinter3   ←   Splinter1
Foxtrot4 = Splinter4   ←   Splinter1
Foxtrot5 = NewMirror2   ←   Splinter1
Foxtrot6 = NewMirror3   ←   Splinter1
Foxtrot7 = NewMirror4   ←   Splinter1
Foxtrot8 = Mirror of Foxtrot1
Foxtrot9 = Mirror of Foxtrot2
Foxtrot10 = Mirror of Foxtrot3
Foxtrot11 = Mirror of Foxtrot4
Foxtrot12 = Mirror of Foxtrot5
Foxtrot13 = Mirror of Foxtrot6
Foxtrot14 = Mirror of Foxtrot7

Golf1 = Splinter2   ←   Splinter2
Golf2 = Splinter3   ←   Splinter2
Golf3 = Splinter4   ←   Splinter2
Golf4 = NewMirror3   ←   Splinter2
Golf5 = NewMirror4   ←   Splinter2
Golf6 = Mirror of Golf1
Golf7 = Mirror of Golf2
Golf8 = Mirror of Golf3
Golf9 = Mirror of Golf4
Golf10 = Mirror of Golf5

Hotel1 = Splinter3   ←   Splinter3
Hotel2 = Splinter4   ←   Splinter3
Hotel3 = NewMirror4   ←   Splinter3
Hotel4 = Mirror of Hotel1
Hotel5 = Mirror of Hotel2
Hotel6 = Mirror of Hotel3

India1 = Splinter4   ←   Splinter4
India2 = Mirror of India1

Results:

Stage #1 + Stage #2 + Stage #3 + Stage #4 = 100 Keys

Elite Afterthought

"Elite speak" is the replacing of numbers for letters and sometimes symbols. Started by hackers, but propagated by gamers, it has become known by the name "l33t speak."

Elite Speak Examples:

A = 4	I = 1	S = $
E = 3	O = 0	T = 7

In the Instant Spy encryption process I have detailed, numbers have to be written out as words, with the exception of the key ID number at the end. Thus, there are conspicuously no numbers. By adding an extra post-encryption programming loop (Elite Loop), the final encryption could be written out in hacker elite speak, that includes numbers and symbols along with the alphabet letters. When decrypted, the elite speak will first be removed by the program, prior to decryption.

Operational Security

The last of the process is the creation of a burn list. The keys will be used in non-sequential order, selected by the software at random. Given this, the burn list will record the code numbers of used keys so they can only be used to encrypt once. They will remain in the codebook database to allow multiple decryptions.

The reason I call it the "Double-O Encryption Process" is that once you have the codebook database, you can place identical copies of the software on two flash drives (Double), which now have the same one-time cipher keys (O). There are several advantages to doing it this way. Although the encryption has been proven to be crack proof, an adversary can still hack your computer to steal the keys. In theory, if the program is run off of the flash drive and removed from the computer after use, it makes hacking or physically stealing the keys very difficult.

Dissociative Behavior

The next barrier to successful application is distribution. If the only way to break the code is to steal the keys, discretion makes sense. If you just want to talk privately with family and friends, slip a flash drive to them at a convenient moment. However, if you are say a government infield operative things get a little more complicated. If you are going to use this product for a professional application, it may be best to use traditional distribution methods from the Cold War. Tradecraft techniques like the "brush pass" and the "dead drop" are good options.

The burn list will keep the two flash drives synchronized, so once they are initially setup, everything will be automatic. If in fact the process has been compromised, a self-destruct system has been built in to wipe out the database. You can never be too careful.

This version of the software has been designed to create a total of 1,000 code keys in 100 key blocks, which are labeled. In theory you could have different blocks of keys for different contacts. In an emergency, one of the contact associated blocks could be self-destructed. This method could allow for a covert network of communication.

Quantum Cracker Killer

Enough chit-chat, let's send a message. Only letters can be used in a one-time pad cipher. This excludes things like punctuation, emoticons, and emojis. Numbers have to be written out as words. The software is designed to allow the user to type naturally, with the only restriction being that you can't exceed 135 letters. Afterwards, everything that isn't a letter is taken out, including spaces. You will be given the ciphertext. You then copy and paste the ciphertext onto the Internet in any way you choose.

Your contact will receive the message and decrypt it. When the flash drive decrypts, it will add the used key number to the burn list. To demonstrate I have provided a sample:

Ciphertext:

DMGRFQU3ZKB0GCCF7NPU37RWZK37Z00UDP0BR071WR$MFKB 44KM01YJKZC3NVFPGN0MBCJN0P7B0VPCJX7P7V7ZWDZLFMYK FV7U7FWVWJNXV$4F1NWD0Q$GRZPP1CY4WQ4ZFMCG-#001

Plaintext:

MEET MIDNIGHT ON CHARLES BRIDGE PRAGUE I WILL BE DRESSED AS CIRCUS CLOWN LOOK FOR BIG FLOPPY SHOES BRING LIST OF RUSSIAN SPIES IN US UNDERSTAND YOU LOOK LIKE MR BEAN

The amount of information you can get into a single 135-character message is very high. This message can be texted, emailed, or tweeted on Twitter. You may be thinking how powerful is the idea of tweeting secret messages on Twitter? Keep in mind that resistance fighters called in NATO airstrikes on Twitter during the overthrow of Muammar Gaddafi.[13]

Video Reveal

When I was done with my project, I submitted a YouTube video that was limited to two minutes demoing the software, a white paper, and a pile of computer code using Python, JavaScript, SQL, CSS, and HTML. The video even featured a CGI Drone attack. I got a perfect score, a cool certificate, and a T-shirt.

Literary Mixology

It occurred to me that there is nothing in my background or experience to really prepare me to write a full-length book. Obviously, I have written a lot of papers in my life, however they have always been concise and to the point. Even in general conversation, I tend to be rather efficient. It occurred to me that I might need to consult a couple of literary masters.

So, to prepare myself, I started reading the classic fiction of Herman Melville and Ian Fleming. Melville's "Moby Dick," has given me some ideas as far as literary execution. His novel is basically a mental download of everything he knows about whales and whaling, while at the same time telling a cautionary tale about obsession. This is conceptually, alarmingly similar to what I am trying to do. Of course, in my book I will metaphorically catch my great white whale. Although regarded as the best book of all American classic literature, "Moby Dick" is very long, and I found some of it rather tedious. It's not like the various movie adaptations. So, to give my writing a little style help, I have read some of Fleming's "James Bond" novels. From my point of view these are just fun books. This is what I want for my book.

So stylistically, between Melville, Fleming, and my personal comedic propensities, this work should have its own unique flavor. This will be literary mixology.

As I mention before, I selected "Moby Dick," because it's about obsession, which is my chief personal flaw. In the book Captain Ahab is said to be a self-proclaimed "madman." Later his monologue is referred to as the "ravings of a madman."[14] I thought that sounded like the right title for my book. So, the original manuscript was named after a line in Moby Dick, because it's a story about an obsessive quest. Only in my version, again, I catch my great white whale!

Listening to Lao-Tzu

This literary mixology was used to successfully create a book manuscript. Upon reading my own completed manuscript, and with the passage of some time, a more powerful ending occurred to me. Rather than just tacking it on to the end of the original manuscript, I decided to use the benefit of a fresh perspective to rewrite the work.

Over the years, I have read a paperback copy of Lao-Tzu's Tao-te Ching several times. It's one of my favorite books. I was exercising and decided to listen to an audiobook version. While I was listening, I started comparing the elaborate first book manuscript I created to Lao-Tzu's powerful book that is both short and simple. I thought what if I applied the principles of the Tao to rewrite and enhance my prior effort.

This version is both a remake and a sequel, as it has a tighter presentation with a new rock star ending. The new title now reflects the stronger concept. For better or worse, I have spent a disproportionate amount of time living in my own head. As such, this is my definitive memoir.

Therapy

I write this book as a form of self-therapy. It's the record of a long journey of trying to understand the world around me. I look at this book as an elaborate journaling exercise. As such it's irrelevant whether or not it's ever published. Like most people, I am just on a journey to make myself feel whole. I have concluded that the best way to seek inner peace is to complete my life's intellectual quest, by finding the answers to my questions, and writing them down. In doing so my journey is complete.

PART 1: PHILOSOPHY

CHAPTER 1: ASIAN PHILOSOPHY

N ow I would like to consult with the great minds of Ancient Asia in the area of philosophy, and discuss Asian influences in my overall thinking. Specifically, I will be focusing on the subjects of nothingness and void.

Chi Deal

First, I would like to discuss the thoughts of the eastern philosopher Lao-Tzu (Laozi). There are several prominent characters in Asian history for which historians have hypothesized as to not have existed. Instead they are said to be amalgamations of many persons and collective wisdom. Unfortunately, Lao-Tzu is one of these persons. However, I am not interested in rewriting history; only learning from it. So, my point of view is that he was a real person. Keep it simple, I say.

Lao-Tzu, who lived roughly between 570 and 490 BC, was a Chinese philosopher and is the founder of Taoism (Daoism). He was born in the province of Ho-nan, China, and there he was a court librarian. According to tradition, he is the author of the Tao-te Ching, which translates to the "Classic of the Way and Its Power,"[15] It's Taoism's fundamental text. Though a small book, it has had an enormous influence on Chinese thought and culture. It focuses on the recognition and acceptance of nothingness as fundamental to understanding the way of the cosmos. Indeed, Lao-Tzu identifies nothingness as the universe's most important component.

I interpret this book to be about how to live in harmony with the positive and negative energies (yin and yang) of the "Tao" or the "the Way." It also talks about political governance, in the context of being in harmony with the "Tao." Independent of the topic, Lao-Tzu consistently favors the weak or the empty over the strong or the full, as he believed that the emptier something is, the closer it's to the true primordial essence of existence.[15]

"The myriad creatures in the world are born from something,
and something from nothing."[16]

- Tao-te Ching

In Taoism there is a concept known as "chi" (sometimes spelled "qi"). What is chi? The Chinese word "chi" means internal energy, which is basically like life force. It's believed that chi energy determines human mental and physical conditions.[17] Chi internal energy can be cultivated through the meditative movement of the art of "Tai Chi," with the express purpose of harmonizing the body's yin and yang energies, and bring it into Tao.

Asthma

Between the ages of five and eighteen, I suffered from severe episodes of asthma. This situation highly curtailed the enjoyment of my public school years. For three years, I took allergy shots and every medication available. In some cases, the medicine was worse than the allergies themselves. For several years a rip-roaring attack became part of my Christmas tradition. That time of the year formed the perfect storm of allergens. It was truly a case of the best of times and the worst of times.

When I went away to college, I had a unique opportunity for an experiment. I could choose whether to take my medicine or not, so I went cold turkey. Also I found this book written by a Tai Chi master. It was "T'ai Chi Ch'uan" by Cheng Man-ch'ing. In it the master/practitioner cured his tuberculosis.[18] I thought, that's respiratory, and much worse than asthma. So for a number of years I did Tai Chi in private using books and videos.

I think it worked. By my early twenties, all the worst symptoms were gone. However, looking at it from a purely scientific viewpoint, there were other factors. During my college years I moved from the polluted Virginia exurbs of Washington, D.C. to the subtropical Central Florida environment of clean air and high humidity. I stopped taking my medicine. There is of course the issue of age. I was old enough to have outgrown my allergies. Personally, I think it's a far more ideal notion that I cured it with Chinese magic.

Now days I still have asthma, but nothing compared to my youth. From time to time, I have a sharp, severe attack, but it only lasts a couple of minutes. On average it only happens a few times a year. It's a far cry from the nearly constant illness of my school days. In my mind I owe my recovery, at least in part, to that "chi deal."

Enlightenment

Siddhartha Gautama, lived roughly between 563 and 483 BC. He is known by most as "the Buddha," which means the "Enlightened One." He was an Indian philosopher and the founder of Buddhism. He is famous for having attained spiritual enlightenment under the famed bodhi tree.[19] In the writings attributed to him, like the Dhammapada, there is no conflict to the views of Lao-Tzu. In fact, in a sutra attributed to the words of the Buddha, he makes a declaration that seems directly in accord with Lao-Tzu.

"All things are no different from emptiness; emptiness is not different from all things. Form is emptiness; emptiness is form."[19]

- The Heart (of Wisdom) Sutra

The Buddha's perspective, and by transference Lao-Tzu's, does not seem to be in conflict with pre-Buddhist Sanskrit texts either. Thus, nothingness and emptiness appear to play a universally fundamental role in the overall Eastern philosophical perspective. At least it seems that way to me.

Zen Thing

Another spiritual innovator was the Indian monk Bodhidharma (482?-539?), sometimes referred to as Tamo (Damo). He is credited with both founding what is predominately referred to today as Zen Buddhism, and the invention of Chinese Kung Fu.

In 518, Bodhidharma, traveled from India to China to spread Buddhism. His travels eventually took him to the now world famous Shaolin temple, a legendary monastery in northern China's Henan province, where his teachings emphasized meditation. He is said to have spent nine years sitting in a cave contemplating a blank wall, consequentially attaining enlightenment.[20]

Many people consider Bodhidharma, as the founder of Kung Fu, and subsequently to be the inventor of all Martial Arts in general. I am aware of some of the nuances of the arguments for and against his direct involvement, and/or the extent of his involvement. My own opinion is that whether he intended to create Kung Fu as the ultimate product of his teachings, I think that Kung Fu as it's now recognized would not exist without Bodhidharma's original teachings. So, it's to him that I am grateful to for the myriad of excellent Martial Arts movies, for which I have wasted many hours of my life watching. As with Martial Arts, Bodhidharma's belief system, originally called Chan Buddhism, successfully found its way to Japan where it acquired its recognized name, Zen Buddhism.

Japanophile

I was once a Japanophile. I am not really sure when it started, one day I just got sucked into the paradox of the ancient and ultra futuristic nature of Japan. I just felt one day that I needed to learn Japanese. It was a hybrid preoccupation, between Martial Arts and Japanese. There was a period where I really liked watching Steven Seagal Aikido Movies, and even certain "heroes in a half shell." My dormmates briefly nicknamed me "Ninja Turtle," because every time they saw me I was either doing Tae Kwon Do or eating pizza. At one point I got recruited out of my Japanese class to take Shotokan Karate. This is when I got deeper into authentic Japanese culture. I would say about 50% of the Shotokan students were actually from Japan. I got to practice my Japanese, when they weren't practicing their English on me. This was a time when I got crazy about all things Japanese, including Japanese women.

The great thing about being a Japanophile in Florida was the food. With local water access, the sushi was spectacular. I didn't really learn about how to order specific types of sushi, because I always ordered "o-makase," or chef's choice. I wasn't always sure what I was eating, but it was always the best.

Now days I am nearly recovered, although I have occasional relapses. I bought a special Japanese headband on the Internet from a store in Japan. It says in Kanji Japanese writing: "Zettai Goukaku," meaning "I will definitely pass the test." It's literally like a Japanese thinking cap for students studying for a big exam. When I need a little extra mental mojo, I wear it. I wore it when I wrote the original version of my "theory of everything" and I wore it when writing this book.

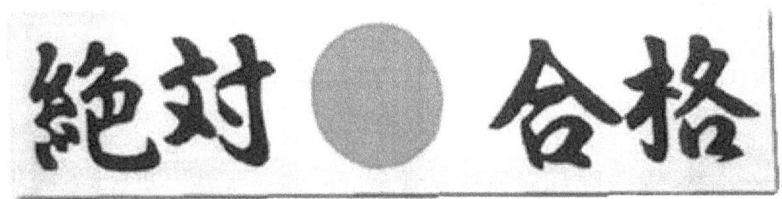

A Picture of My "Deep Thinking" Headband

I'm currently pursuing MMA-based fitness, or Mixed Martial Arts. However, this time it's less for self-defense, and more for my quest to downsize myself. I think MMA is very much in the spirit of Bruce Lee's book the "Tao of Jeet Kune Do." In this classic Martial Arts manual, he describes a mix of fighting styles that were traditionally practiced separately. I really like the idea of MMA because it's like a grand unification theory for Martial Arts.

Zen

Bodhidharma's Zen is unique, as it's regarded as an intuitive school of Buddhist meditation, in that it has nothing to teach and no rituals. It's a method of self-training that leads to an understanding of reality. The basic idea is that the individual can discipline their mind so that they can come into contact with the inner workings of their being, and intuitively grasp what can't be grasped rationally. The Zen practitioner believes that this "awareness" can't be taught, and that each person must find it for themselves.[21] Zen Buddhist thinking combines elements from both Buddhism and Taoism traditions; as such it's definitely not in conflict with their shared emphasis in nothingness and emptiness.

Zen Buddhism is important to me because I was raised Zen Buddhist by my parents. Although I tend to look at the world from a secular viewpoint, at heart I am a Zen Buddhist. I think this is at the root of my open-mindedness.

Mirror

The title of the first patriarch of Zen is attributed to Bodhidharma. An example of Zen might be an exchange between the fifth and sixth patriarchs of Zen. The fifth patriarch, Hung-jen, announced that it was time to appoint a successor. Subsequently, he asked the monks to write a poem expressing their understanding of Zen. The head monk wrote (writing on a wall for all to see):

"Our body is the bodhi tree,
And our mind a mirror bright.
Carefully we wipe them hour by hour,
And let no dust alight."[22]

Hui-neng (638-713), one of the guys that worked in the kitchen wrote back:

"There is no bodhi tree,
Nor stand of mirror bright.
Since all is void,
Where can the dust alight?"[22]

Consequently, Hui-neng got the job of the sixth patriarch, having attained the proper understanding of Zen.[22] This is a very good example of, "what is Zen?"

I believe it can be said that there is an overall consensus in the Eastern philosophical worldview as to the predominate importance of the concepts of nothingness and emptiness (also referred to as voidness) in regard to assembling an understanding of the perceivable universe. This is the main point to take away from this chapter.

"Do not seek to follow in the footsteps of the men of old;
seek what they sought."[22]

- Basho (Renowned Zen Monk)

CHAPTER 2: ASIAN ANTIQUITY

In order to truly understand Asian philosophy, it's worthwhile to take a look at Asian history. Ancient Chinese history is the best way to do this. Chinese history is so long and continuous that the earliest periods of Chinese history and pre-history are often regarded as purely mythological. I approached this topic by studying both Chinese history and Chinese architecture.

Flood Myth

Chinese pre-history starts during the Neolithic period following the last ice age. Chinese history starts with a single significant event. Believe it or not this event is Noah's flood. Honest, I am not kidding. The story is there was a great flood. The peoples of what would be China climbed to the heights of the Himalayan mountains. Many animals also survived by running up into the mountains.[23]

Ararat Analysis

If you suspend any skepticism in regard to the premise of the Great Flood myth, and take it on face value, you can reach some interesting conclusions through deductive reasoning. Given that based on the legend that there were Great Flood survivors in China, I thought to myself, I wonder if it could be possible for there to have been survivors elsewhere on Earth. As such I did some informal research via a Google search. I discovered that there are 271 mountains in the world taller than Mt. Ararat at 5,137 m (16,854 ft.), which according to the Bible was initially completely submerged underwater.[24] The tallest mountain in the world is Mt. Everest at 8,848 m (29,029 ft.), which is 58% taller than Mt. Ararat.

So to figure out other possible safe locations, I analyzed it continent by continent. In Africa, you could have survived if you were in Kenya or Tanzania. If you were in Australia (Oceania), or for some reason Antarctica, there would have been no hope. Asia has the greatest number of possible safe havens on the planet. Places one could survive in Asia include: Afghanistan, Bhutan, China, Georgia, India, Iran, Kyrgyzstan, Myanmar, Nepal, Pakistan, Tajikistan, and Tibet. Europe has more of a bleak outlook. The tallest mountain in the Alps is Mt. Blanc at 4,810 m (15,781 ft.), which is significantly shorter than Mt. Ararat. There is a mountain tall enough in the European part of Russia, however. In North America, the only place in the United States that would have had a refuge would be Alaska. There are also mountains high enough in both Canada and Mexico. In South America, the places one could survive include: Argentina, Bolivia, Chile, Colombia, Ecuador, and Peru. However, I may have digressed from the original topic.

Flood Aftermath

According to legend when the waters receded there were very bad problems. The animals survived in far greater numbers than the people. The high places were covered with dangerous animals and the low places were still flooded and didn't drain back into the sea. Also, the trapped waters were said to be filled with dangerous monsters, of which were many snakes. At one point it became clear, with animals outnumbering the people, and all of their infrastructure destroyed, that this was the end of all of known humanity. However, extinction was averted by the innovation of a single leader.[23]

The legendary Nest-Dweller had the idea the people would stop living on the ground and live up in the trees like birds. And so, they built treehouses. It was only for this reason that all the people were not eaten. Over time the treehouses evolved into completely manmade structures. The tree trunks were replaced by columns and the weight of these elevated buildings were held up by these complex devices called dougongs. Dougong literally translates to "cap and block." They are a structural element of interlocking wooden brackets.[23]

Xia Myth

According to writings attributed to Confucius, the world flood happened during the reign of the Xia dynasty sage king Yao. Birds and beasts proliferated, and traditional farming was very difficult. The birds and the beasts "crowded in on the people." The sage king Shun was the successor of Yao, and he employed Yi to set fire to the mountains and the marshes to burn out the animals. As such the animals went into hiding. The successor of Shun was the sage king Yu, who reengineered the nine rivers of China to drain the water from the deluge into the sea. Once again, the people of China could farm.[25]

I was taught in Harvard ChinaX, that the Xia dynasty might not be real and that they were only mentioned so that the Confucian textual references would make sense. I was told that there was no physical evidence for the Xia at all. Imagine my surprise when my Tsinghua Chinese architecture Prof. He Congrong showed a picture of the Yangshao village site, Jiangzhai in Lintong, Shaanxi Province. It was presented as actual physical evidence of the Xia dynasty![23]

Chinese Architecture

The modern form of Chinese traditional architecture can draw its origin directly to adaptations required to survive the environmental adversities immediately following the great flood. With the water monsters gone and the land being farmable again, the people could come down from their nests and the world could begin once again. The dougongs were now used to support roofs instead of elevated shelters. So, you have columns creating a framed structure with the dougongs supporting the roof. All of this was set up on a platform. As a consequence, all outer and inner walls were nonloadbearing. As such the outer walls were often weak and not very water resilient. The solution was to create massive dougong supported roofs with flying eves. This is a distinctive part of traditional architecture to this day.[23]

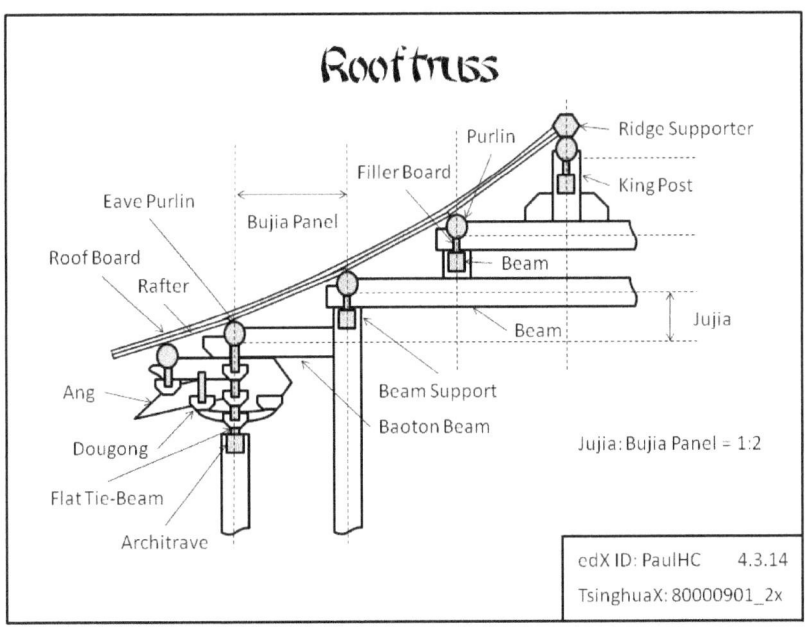

Traditional Chinese rooftruss featuring dougong support structure.

Today you still see all of these elements in traditional Chinese architecture. I learned by reading "A Pictorial History of Chinese Architecture" by Liang Ssu-ch'eng, that in fact the temple architecture of Kyoto, Japan actually represents the premiere Chinese architecture of the Tang dynasty.[23]

The architecture of the Forbidden City in Beijing, China is one-stop shopping for the ultimate examples of Ming and Qing dynasty Chinese architecture. Halls, temples, gardens, it has it all. The roofs and the flying eves are smaller because they are less necessary and the dougongs are very complex and mostly decorative. However, you can look at all the elements in the Forbidden City structurally and technologically, and trace them all the way back to the Nest-Dweller.[23]

Warring States

After the Xia dynasty, was the Shang dynasty. The Shang ruled with the philosophical legitimacy of the "mandate of heaven." They were eventually overthrown by the Zhou dynasty. The philosophical legitimacy of the Zhou was that the Shang had lost the mandate of heaven by being unjust. As such, heaven now apparently favored the Zhou. The Zhou dynasty fell into decline, which plunged China into a long-term civil war between the states of China.[25]

During this time of chaos and uncertainty, almost all of the great literary works of China were written. These included Lao-Tzu's Tao-te Ching, the writings of Confucius, Sun Tzu's Art of War, and the Book of Changes.[25]

Book of Changes

My understanding of the Book of Changes, called the I Ching, is that it's ultimately a book of divination. This divination is based on natural patterns. The nomenclature of the system uses hexagrams, composed of two trigrams. These trigrams can be equal to multiple states. The states are yin, yang, and combinations of both yin and yang.[25]

My personal multidisciplinary approach to life has led me to make an interesting observation. In computer science, conventional computation is done using a binary system. A unit of information is called a bit, and it can have a value of either a 0 or a 1. However, the newest type of upcoming computer is called a quantum computer. It uses a quantum bit. The value can be 0, 1, or a "superposition" of the two. As such, a quantum bit, or qubit, can be equal to multiple states. From a conceptual point of view, the I Ching uses the idea of quantum superpositions for the purpose of calculative divination.

Cosmic Resonance

The Warring States period ended with the state of Qin conquering all of China. The Qin king was declared the first sovereign emperor of China. Emperor Qin not only united China into a singular country, but also united Chinese into a singular written language. The philosophical legitimacy of the Qin dynasty was not the mandate of heaven, but a concept put forward called "cosmic resonance." I believe the principal takeaway from the cosmic resonance theory is that the incredible chi power of the emperor would set the nation in order.[26]

Actually examining cosmic resonance is very strange. It says some really crazy stuff. It would be completely reasonable to imagine a Star Wars character saying some of these things. A modern person could be forgiven for thinking that this work might be describing "the Force."

So, the theory goes that the phenomena of human consciousness and cosmic resonance both emerge from the dynamic properties of chi. As such one can use one's consciousness, given enough chi, to achieve ESP through cosmic resonance. This implies being able to feel real-time events at great distances.[26] This is equivalent to a Star Wars-like "disturbance in the Force." Also, it gives one the ability to foresee impending future dangers. This is equivalent to the Star Wars catchphrase, "I have a bad feeling about this."

It discusses concepts like the idea of probabilistic interference between two unrelated concurrent events in near proximity. Basically, "a theory of simultaneous, non-linear causality."[26] This reminds me of an aspect of quantum mechanics called quantum decoherence. It's difficult to isolate a quantum system, because information can leak in from the environment that changes the system.

Actually, there is a lot about cosmic resonance that reminds me of modern physics. A sort of vanguard theory of modern physics is that of zero-point energy. Simply stated, it's continually revitalizing energy that exists in the empty space of the universe. It's often referred to as vacuum energy.

In cosmic resonance theory, it's said that chi "composes all objects in the world and fills all spaces between them." Chi is the animating force that separates the alive from the dead. Chi fills all space and conducts "sympathetic vibrations" between objects.[26] Some of this sounds very close to the current understanding of vacuum energy. Also, if you change the words around a little, it sounds like the Star Wars character Obi-Wan Kenobi's explanation of "the Force" to Luke Skywalker.

If a person looked at this theory without the benefit of historical context, it would be reasonable to say that someone simply studied the works of Albert Einstein and created a bizarre philosophical text to justify a political goal. Of course, with the benefit of historical context, it's obvious this text was presented around the time of the unification of China, in 221 BC.

Cyclic Forgetfulness

What is the source of this pre-Einstein era information? My opinion is that it's the residual knowledge of a previous pre-flood human world, that was preserved by the bravery of people holding onto the tops of the Himalayan mountains. If human civilization is getting destroyed over and over, perhaps there was an earlier version that was our technological superior. If this hypothesis is true, this type of knowledge would most likely have to popup somewhere else in the world, in other wisdom traditions.

CHAPTER 3: MODERN THINKERS

Modern thinking more or less begins with the scientific revolution. The scientific revolution started during the period known as the Renaissance. The Renaissance was a period of great innovation and growth in the arts. As such, a unique way of gaining insight to the interworking of the scientific revolution might be to look at what was happening concurrently in the art world. Specifically, in the world of the Renaissance painters of Europe.

Painterly Perspective

There are many models for analysis. I personally lean toward the contemplation of art works, as opposed to a detailed study of artists or historical facts. In this way, you can see art from the actual viewpoint of an artist as opposed to the intellectual separation of students solely memorizing dates and periods. I think it will never be possible, however, to fully separate the biases of the present day in looking at art from the past. The fault is that we are attracted to art that speaks to us, and this art has a lot to do with our contemporary point of view.

Over the centuries there have been many theories and methods for looking at the history of art. From antiquity, the Greeks and Romans have given us a progressive model of looking at art as a less to more process, as in Pliny the Elder's book "Natural History." During the Renaissance, a biological approach was taken that saw art as a single organism going through the cycles of life, as described in Giorgio Vasari's book "The Lives of the Most Excellent Italian Architects, Painters, and Sculptors" (1550). The connoisseurship model of the seventeenth century focuses more on the style and individuality of the art. Art has also been studied from the perspectives of evolution, form, and historical narratives. The modern model of deconstruction sees the linking of art to a specific narrative as an imperfect analysis.

After taking an edX art course given by Universidad Carlos III de Madrid, I see art as a type of Darwinian evolution, and that the works can't fully be understood separately. It's clear to me that there is a type of collective learning taking place. I have the bias of my own time, as I see a reflection of my own "copy and paste" culture in the evolution of these paintings. Artists studied other artists, and then incorporated the parts they liked into their own works. There was a lot of copying and remixing of ideas. In this way, there's a flowing continuum of group thought. By looking at art history in this way, one can plug into the logic of the process, and experience the art as an insider, as opposed to the outsider looking in.

It's fairly easy to give specific examples of the concept of the flow of ideas from one artist to the next. Take for instance Caravaggio's style of chiaroscuro, seen in paintings like "Crucifixion of Saint Peter" (1600-1601) and "The Inspiration of Saint Matthew" (1602-1603). Chiaroscuro reappears in the works of Jusepe de Ribera's "Saint Andrew" (1631) and in "Martyrdom of Saint Andrew" (1638). Even later, chiaroscuro is then seen again in Mattia Preti's "Saint Sebastian" (1657).[27]

Another way ideas are transmitted is not only in style, but also in content. In Joachim Patinir's "Charon Crossing the River Styx" (1520-1524), he incorporates elements of the work of Hieronymus Bosch, like those found in the triptych "The Garden of Earthly Delights" (1500).[27]

I see parallels between this artistic process of collective learning and the development of science. In science there is also a lot of copying and remixing of ideas. So just like in the art world, science is a flowing continuum of group thought.

Albert Einstein

The first person I would like to introduce is Albert Einstein (1879-1955). Albert Einstein's work built upon the previous works of scientists such as Sir Isaac Newton and James Maxwell. He created the special theory of relativity, which later evolved into the general theory of relativity. These successful theories of Einstein essentially comprise almost all of modern physics to this day.[28]

"In 1915 a completely new mathematical model was put forward by Einstein: the general theory of relativity. In the years since Einstein's paper, we have added a few ribbons and bows, but our model of time and space is still based on what Einstein proposed."[28]

– Stephen Hawking

Of course, he also created the theory of quantum mechanics, which makes up the remainder of present-day modern physics. So essentially, Einstein is modern physics. To my knowledge, it's Einstein that started the quest for the "theory of everything." He dedicated the last thirty years of his life to it.[29]

"The cosmological constant was my greatest mistake."[29]

- Albert Einstein

Einstein believed that the cosmological constant was the essential factor in putting together a grand unification theory. This would be a solution where the present universe could be modeled and the past and future of the universe could be calculated. However, Einstein's starting premise was that time was infinite in both directions, past and future, and that the current universe was static. Thus, he conceptualized the cosmological constant as a balancing agent to make his preconceptions balance out. This motivation was counterbalanced by the fact that general relativity predicts that time should have a beginning and an end. After Edwin Hubble's discovery of the expansion of the universe, Einstein rejected the cosmological constant as a mistake.[28]

Carlos Castaneda

The next person I would like to introduce is Carlos Castaneda (1925-1998). Though many would probably question his work (for many reasons), he is a legitimate anthropologist, who reported from a scientific perspective on the ancient Toltec culture of Mexico.

Don Carlos' books have had a major effect on modern thinking, and they have revealed many lost pieces of human knowledge to a modern audience. For example, Carlos Castaneda's works have influenced the teachings of Prof. Joseph Campbell, who strongly influenced George Lucas, who in turn created the "Star Wars" Saga. Which strongly influenced me.

He wrote a string of bestselling books from 1968 to 1998, starting with "The Teachings of Don Juan: A Yaqui Way of Knowledge." As an anthropologist, Carlos Castaneda imbedded himself in a native culture, becoming a member of that culture, and then reported the internal workings. Carlos trained for and became a Yaqui shaman or sorcerer or brujo, one of the practitioners of the perhaps dying Nagual tradition, based on the ancient Toltec wisdom tradition. In doing so he exposed the modern Western world to new conceptions of reality.

In 2012, I audited an Oxford University course on quantum mechanics on iTunes U. It was taught by Prof. J.J. Binney, and the whole thing was roughly 20 hours long. I was alarmed and amused by the fact that a lot of the advanced concepts that were presented correlated 1:1 with stuff in the Carlos Castaneda books!

Stephen Hawking

The last person I would like to introduce is Stephen Hawking (1942-2018). He is popularly known to most people by way of the 1988 bestselling book "A Brief History of Time" and the 2014 movie "The Theory of Everything." In the way of Carl Sagan, he popularized science to a mass audience.

Steven Hawking is famous for his theoretical contribution to work on black holes, in collaboration with Sir Roger Penrose. He also was distinguished by his personal battle with a rare form of amyotrophic lateral sclerosis (ALS). He communicated with the world through a speech-generating device connected to his wheelchair.

Like Einstein before him, he searched, without results, for the theory of everything. As such, the task remains to be completed.

PART 2: AEROSPACE

CHAPTER 4: ROCKET REALITY

When I was college age, I studied some aerospace engineering. I had the unique opportunity to refresh my skills through edX and MITx. MIT Prof. Jeffrey Hoffman starts his astronautics course with something he calls the "Aerospace Perspective." I was so impressed with this, I decided to steal a little bit of it for the introduction of this chapter.

Aerospace Perspective

Imagine that the Earth is only 1 m in radius, which is the size of the model in the Museum of Science in Boston. Mt. Everest would only be 1 mm above sea level, and Commercial Jetliners would be only 2 mm above sea level.[30]

Other than Apollo, all human space activity has taken place in the exosphere layer of the Earth's atmosphere. The official definition of space is the Von Karman Line, which is 100 km (62 miles) above sea level. According to the scale of the model, you would only have to travel 1 ½ cm to receive your astronaut wings. This vertical distance is closer than New York City to Boston or London to Paris.[30]

The lowest ever Space Shuttle orbit was 200 km, and the International Space Station (ISS) is in low Earth orbit (LEO) at 400 km. To continue the thought experiment of a 1 m Earth, ISS would be at 6 cm above sea level, which is about the width of a hand. Geosynchronous Earth orbit (GEO) would only be 5 m.[30]

The farthest traveled by humans is 400,000 km to the Moon. According to the model this is equal to 60 m (200 ft.) Only 24 people have made the journey to the Moon, and of those only 12 have walked on it. Mars is 40,000,000 km away, or 6 ½ km relative to the model.[30]

Voyager 1 is 3,000 km away relative to the model, which is the same as Boston to Denver. It's currently in what is considered interstellar space.

Proxima Centauri, the nearest star, is 40,000,000,000,000 km or 4 light-years, which relative to the model would be 6 million km. As such, Voyager 1 would take over 100,000 years to reach Proxima Centauri, if it were headed in the right direction, which I believe it isn't.[30]

Our Sun in the Milky Way is 2/3 of the way out from galactic center to galactic rim. The Andromeda galaxy, the nearest major galaxy to the Milky Way, is 2 million light-years away, and is regarded as the farthest object visible to the human eye.[30]

It's important to look at astronautics in perspective. The big picture is that humans are currently still at the earliest stage of space travel, and there is plenty of room for improvement. If you understand the "Aerospace Perspective," it's ego deflating.

Wright Brothers

It's important to start at the beginning. Mankind's heavier-than-air ascension into the sky starts with the Wright Brothers. This was the birth of aerospace engineering. Rather than attempt to regurgitate the history around the spectacular accomplishment achieved at Kitty Hawk, I think it might be wiser to focus on what I have personally taken away from this real world lesson. I learned that no dream is too big, and no one is too small to succeed. The airplane was not invented by a government research project, a faceless major multinational corporation, or even by college graduates. It was done by two bicycle repairmen.

You could say well that was a long time ago, the little guy, acting independently, couldn't possibly amount to much now. But take the modern examples like Bill Gates (Microsoft), Steve Jobs (Apple), Mark Zuckerberg (Facebook), Sir Richard Branson (Virgin Galactic), Elon Musk (Tesla/SpaceX), and Jeff Bezos (Amazon/Blue Origin). The next great thing is more likely than not to come from another little guy, with a big idea.

Rocket Truth

The space age officially began when rocket power put the first satellite into orbit. The launch of Sputnik in 1958 began the Space Race between the United States and the Soviet Union. The Space Race concluded in 1969 when rocket power put two American astronauts on the surface of the moon.

However, since the Apollo era, which resulted in one of mankind's greatest achievements, human exploration has essentially stopped. As of late, all human experience has been limited to orbital achievements, such as space shuttle missions, and space station activities. Exploration itself has not stopped. Space telescopes and robotic exploration craft have taken over.

A question that has intrigued me is why? Why did human exploration essentially stop? As a child, school provided lots of answers, such as "we could continue if we wanted to, but there is no political will" or "it would cost too much" or "that money is better spent down here, for socially responsible programs." I believe, however that these reasons are false. The idea that we simply could continue human exploration if we wanted to, but we don't feel like it right now, is to make ourselves feel better. The real truth is that we can't do it right now, period.

Okay, so why can't we do it. Because, all are major advances so far have relied on rocket power. Conventional rocket power can't put an astronaut on Mars. Thus, our flag planting days are over until we come up with something better.

What is so bad about conventional rocket technology? Well consider the following. Assuming you start on the Earth's surface (which is a reasonable assumption), you have to increase the amount of fuel relative to the altitude you want to achieve. But this of course works against you, as it increases the weight of the vehicle you are trying to lift. The solution is that rockets are built in stages, such that as the fuel in one stage is consumed, it's discarded to offset some of the weight penalty. Ultimately, the problem is that you can't have an infinite number of stages. So, what is the problem really? Is it will power? No, it's just math.[31]

- Stage #1:   3 km/s
- Stage #2:   7 km/s
- Stage #3:   11 km/s

One main stage will take a crew vehicle sub-orbital. Two main stages will take a crew vehicle into orbit. Three main stages take a crew vehicle to the Moon. Four main stages will be too heavy to get off the ground.[31] Herein lies the problem. A crewed mission to Mars awaits propulsion technology that is far superior to rocket technology.

In the absence of an alternative to rocket power, there is only one workaround. If you lighten up the payload from a crew vehicle to a robotic exploration vehicle, only then can you go to Mars or any other location in the solar system. Robotic exploration isn't better than human exploration; it's just lighter in weight. It's time for a better choice. Unless we find something better than conventional rocket technology, the human exploration stops here.

Prof. Hoffman did an interesting calculation. He proposed the question, "How much energy does it really take to get 1 kg to orbit?"

$$
\begin{aligned}
\text{Total Energy} \quad &= \text{PE} + \text{KE} \\
&= 3.6 \times 10^7 \text{ Joules} \\
&= 3.6 \times 10^7 \text{ Watt-sec} \\
&= 10 \text{ Kilowatt-hrs}
\end{aligned}
$$

The cost of 1 kW-hr from the NSTAR Electric Ion Engine equals less than $1. Thus, it should cost less than a $10 to put 1 kg into orbit. Today it typically costs $10,000 to $20,000/kg. According to Prof. Hoffman, "We can do better!!!"[30]

SpaceX Starship

The SpaceX Starship is designed specifically to make humans a multiplanetary species.[32] I think it's important to point out that "Starship" is not a starship, which by definition is a crewed vehicle capable of interstellar, not interplanetary travel. It's just called Starship because it sounds cool, which it does.

Starship seems to contradict my previous argument about the limitations of rocketry. My position is that it really doesn't. Starship uses 21st century technology to rework the problem, without changing the underlying limitations.

To the casual space enthusiast, Elon Musk seems to propose a two-stage luxury direct spaceflight to Mars. But, I previously stated that it would take four stages. Well, the Starship is a two-stage rocket that will takeoff and place the Starship spacecraft into orbit. Then the first stage will return to Earth, where it will be refueled and reused. A Starship tanker spacecraft will be added to the top of the refueled Super Heavy booster rocket first stage, to create yet another two-stage rocket. The Starship tanker spacecraft will dock with the Starship spacecraft and transfer fuel. The refueled Starship spacecraft will leave the Earth and go to Mars.[32] So ultimately, if you count it up, it will still be four stages.

The Starship's heavy lifting magic could be of great value in an aggressive return to the Moon. Elon Musk's proposal of creating a full-scale Moon base sounds really cool. The aspiration of returning the world back to a time of exploration is admirable.

The concept of a crewed Mars bound Starship spacecraft is at best ambiguous, and at worst a little bit sketchy. What is clear is that it can't go any faster than the limitations of chemical rocketry. The Starship spacecraft is big. Big and slow is better than small and slow, as at least it has the size to add a lot of radiation shielding.

Although I'm still skeptical about certain aspects of the plans of SpaceX, I have become an enthusiastic fan. I regularly watch the launch webcasts on YouTube.

Pad 39A

When I was a student member of AIAA, I went on a VIP tour of the NASA Kennedy Space Center. The pinnacle of the trip was going to Pad 39A. At the time it was the launch pad for the Space Shuttle. That day the pad was not in use, so we could stand directly on it. It was a clear blue day. Rather than just looking around, I stood there and looked straight up. I knew this is where Apollo 11 took off from. I wanted to see the same patch of sky that they saw. This was the spot where it happened!

Presently it's the home of SpaceX's effort to return to the Moon, and maybe even launch a crewed mission to Mars. I will always remember my moment on Pad 39A. As such, I feel a weird spiritual connection with SpaceX's endeavors.

Replacing Rockets

In response to the need to get away from conventional rocketry, there have been many false starts to real, non-robotic space travel. For the most part it's a series of ideas from the best and the brightest the world has to offer. Most of them are terrible.

False Starts

Catalogue of Crap:[33]

- Antimatter Rockets
- Catalytic Ramjet
- Dean Drive
- Electromagnet Launchers
- Icarus-1: A Solar Sail Interstellar Probe
- Interstellar Laser Ramjet
- Interstellar Laser Rocket
- Nuclear Rockets
- Pellet Stream Propulsion
- Quantum Ramjet
- RAIR: The Ram-Augmented Interstellar Rocket
- Sark-1: A Solar Sail Interstellar Ark
- The Matloff/Fennelly Electromagnetic Ion Scoop
- The Whitmire Electromagnetic Ion Scoop

"A" List

Omitted from this list are few projects I would like to look at in more detail. The following are the best concepts created so far. Out of respect please refrain from laughter.

Daedalus:

This design from the British Interplanetary Society is setup to have a 10% speed-of-light maximum velocity. The design is completely contingent on the use of a nuclear fusion reactor, which is much safer and efficient than existing fission power plants. Of course, nuclear fusion reactors don't exist. This is a serious problem.[34]

There have been some notable fusion breakthroughs in 2021 and 2022, however a usable reactor is still a long way from providing useful energy. At this point a reaction can't be sustained for greater than 5 seconds. So, for now, and the foreseeable future, fusion is not an option.[51]

Bussard Ramjet:

This concept, unlike the first, is designed for near speed-of-light travel. It's designed to scoop up diffuse matter, dominantly hydrogen atoms, which are floating between the stars, and accelerate it into a fusion engine and eject it out the back. Hydrogen would be both the fuel and the reaction mass.

Once again, fusion reactors don't exist. Also, in deep space, there is only about one atom in every ten cubic centimeters. This means that for the ramjet to work, it will need a frontal scoop hundreds of kilometers across.[34]

Ion Propulsion:

This concept holds a unique place, as it's the only design in this chapter that has actually been built and flown by NASA. The technology sacrifices high thrust for high exhaust velocity. An onboard electric generator ionizes atoms and powers an accelerating electric field. The electric field accelerates the ions to a high directed velocity. Electrons stripped-off during the ionizing process are sent rearward to neutralize the charge in the beam.

The acceleration, though continuous to a high velocity, is nothing short of laughable. The thrust of NASA's DS-1 barely produced enough thrust to move a sheet of paper. It's okay for select applications, but it's almost non-existent payload capacity removes it as a candidate for crewed space travel. Ultimately, it's a sad and lame technology.[33]

Big Mistake

Every technology I have mentioned has either been unusable or laughable. There, however, is a lone exception. It of course has been summarily disqualified, and forgotten.

Project Orion:

I first learned about the Orion watching Carl Sagan's "Cosmos" TV series. This spacecraft is designed around the concept of using the explosions of nuclear bombs to push against an inertial plate. Each explosion would provide a "putt-putt" effect, like a nuclear space motorboat. This design is intended to produce a non-relativistic starship, with a 10% speed-of-light maximum velocity. What makes the concept unique among all that are listed in this chapter is that it's the only high-performance design that could be built now.

This nuclear weapon based design will release vast quantities of radioactive debris. Additionally, there is an international treaty that forbids the detonation of nuclear weapons in space. Thus, technically the Orion would be "illegal" for space travel.[34]

So, the only working design for human deep space travel was banned. Yeah, that's the world I know. The abandonment of Project Orion is a tragedy that is difficult to measure.

CHAPTER 5: REAL STARSHIPS

After a lifetime of thinking about the problem, and comparing notes with smarter, well qualified people, I have come up with what I humbly consider to be a solution. My original concept, the "Electron Propulsion Engine," satisfies the optimum criteria for a realistic starship, and such a vehicle lends itself to wild applications.

Electron Propulsion

Performance:

Electron Propulsion Technology (EPT) produces a 10 million Newton constant thrust, with a non-prohibitive power plant mass. The most important point about EPT is that electrons are being accelerated in packets of 1 C (Coulomb), which is dissimilar to all forms of electric propulsion known to date.[35] As such it yields a different outcome.

Mechanical Embodiment:

This is a quick view of what physically comprises an Electron Propulsion Engine, which is the machinery of EPT.[36]

- Electron Collection Surface
  157 m^2 Equals 1 C/s in LEO (Low Earth Orbit)
- Electron Injector
  Converts Electric Current into Free-Electron Particle Packets
- Linac Accelerator (Linear Induction Accelerator)
  Length of 1 m

- On-Board Xenon Gas
  Requires 400 kg of Xe per Month
- Quadrupole Magnets
  4-Pole Magnetic Fields Which Keep the Electron Bunches Focused
- Thermionic Fission Cells: On-Board Power Sources
  96% Enriched ^{235}U, NaK Coolant; 7-10 Years Full-Power Life
- Non-Prohibitive @ 10 kg/kW

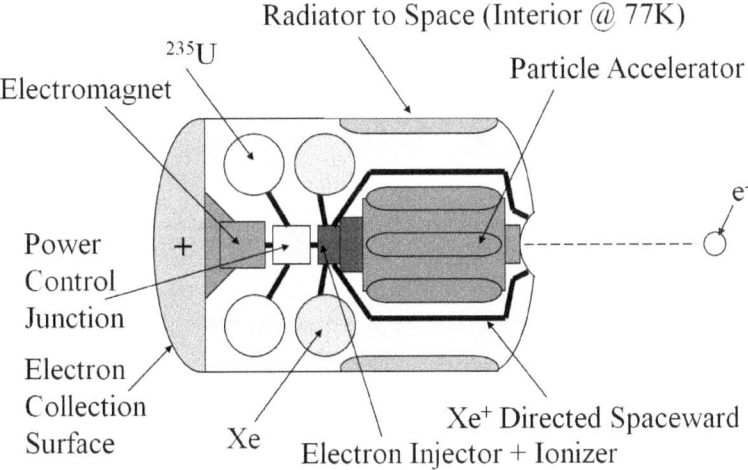

Parts of EPT Engine

Thrust:

I have never liked doing comparisons with ion propulsion, however it's the simplest way to illustrate the performance opportunity of EPT. Given that it's unavoidable, the following is an accurate description of NASA's Deep Space-1 probe thrust.[37]

$$q = \dot{q}t = (1\,C/s)(10^{-7}s) = 10^{-7}C$$

$$F = qE = (10^{-7}C)(10^{6}\,N/C) = 10^{-1}N$$

My engine is a novel approach to well established physics. In accordance with Newton's second law of motion, if an electron's finite mass of about 9.1×10^{-31} kg is to be accelerated, it will require a force. Newton's third law of motion states that forces must always exist in pairs, which are equal in magnitude and opposite in direction.[38]

As stated before the electrons will be accelerated in packets, not individually. At a scale of 1 C per packet, new properties emerge. Causal is the Child-Langmuir relation below (where I is current and V is voltage):[37]

$$I \propto V^{3/2}$$

In the presents of an accelerating field of strength of 9.8x10[6] N/C "the periphery of the non-neutral EPT electron packet 'cloud' will shield the bulk of the electrons from the accelerating field."[37] An electrostatic field is generated by the expanding 1 C packet.[39] This "is an internal symmetric field," so it can't oppose the external field.[39] I have concluded that the resulting acceleration should be near 0 m/s[2] on the 1 C packet, but not equal to zero. However, as stated above, the packet will be expanding, rounding by order of magnitude, at a rate of 100 m/s.[39]

4-pole magnetic fields can "keep the electron bunch focused as it progresses in its acceleration," also referred to as quadrupole magnets, thus controlling the electrons' repulsive force.[40] The electrons will follow a linear path through the linac accelerator, until they are exhausted to space via an exhaust port. With an accelerator length of 1 m, and a travel rate of 100 m/s, a 1 C electron packet will be in instantaneous residence for a period of 1/100 of a second. Thus, 1 C can be maintained in continuous instantaneous residence by injecting 100 C/s.

It has been verified that independent of the resultant particle packet acceleration, that the force placed on the packet can only be defined by the equation F=qE, F (force) = q (total instantaneous charge) times E (electric field).[37] Thus you get the following:

$$q = \dot{q}t = (10^2 \, C/s)(10^{-2} s) = 1C$$

$$F = qE = (1C)(10^7 \, N/C) = 10^7 N$$

If the vehicle is placing a force on the particle packet, F=qE, then an absolutely equal force must be applied to the vehicle. I have already proven that the "action" force of the action-reaction pair will equal 10 million Newtons of force. Thus, if this force is not reacted by 10 million Newtons of force, it will be in violation of Newton's third law of motion.

Restated: Sir Isaac Newton's third law of motion states that forces must always exist in pairs, which are equal in magnitude and opposite in direction.[38] This is not open to interpretation! Any "action" force must be "reacted" by the vehicle. This is clearly the most powerful in-space propulsion system ever realistically proposed.

Fuel:

The engine is designed to operate in two different environments: orbital space (LEO) and interplanetary space. The difference between one and the other is the external ambient electron density.

Given that a LEO particle density (of electrons) is around 10^{13} m^{-3}, at a particle velocity of 7,800 m/s, "there exists sufficient ambient electron flux for EPT to operate at 1 C/s in LEO space."[37] The collection surface will use an applied magnetic field to funnel particles toward the spacecraft.[36] The magnetic field will be able to "suck" them in from a number of meters from the spacecraft.[37] External collection can be continued until the vehicle reaches Earth's "escape velocity," which is about 10 km/s. If you took a hemispherical dome, described by the equation $A = 1/2(4\pi r^2)$, you could create a frontal collecting surface with an area of 157 m^2, with a diameter of 10 m. The "10 m diameter surface is doable" with today's deployable structures.[37]

Note, that for a 100 C/s vehicle, there is only 1/100 of the "fuel" needed to have a sizable craft operate at maximum thrust. However, there is enough for adequate partial thrust (while in the Earth's magnetosphere).

Given interplanetary space electron density hovers around 5×10^6 m^{-3}, with a particle velocity of 400 km/s, for the inner solar system, "there is insufficient ambient electron flux for EPT to operate at 1 C/s in interplanetary space."[37] However, in this environment an internal electron "fuel" source is applicable, thus low electron particle density doesn't prevent the operation of EPT. The idea is to ionize xenon gas, which will be stored on the spacecraft, using power from the thermionic fission cells, which is the engine's internal power source. The freed electrons will be used by the electron injector and the ions will be applied to other work. With an output of 100 C/s, I believe it will only require 400 kg of Xe per month. This is based on the knowledge the Xe has a mass of 2×10^{-25} kg and a charge of 1.6×10^{-19} C per particle.

Spacecraft Charging:

My spacecraft neutralization plan for LEO, where the required electrons for propellant can be acquired from the outside environment, is quite simple. If the same number of electrons enter and leave the spacecraft, concurrently, then no spacecraft charging will occur (equal electrons in + equal electrons out = spacecraft neutrality).[37]

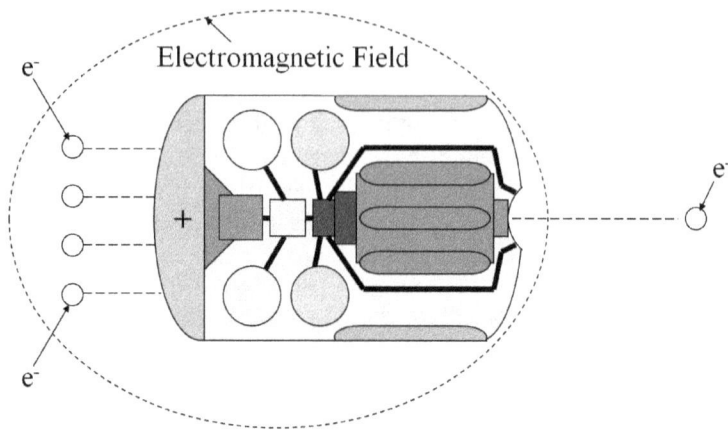

Operation in LEO (Low Earth Orbit)

My spacecraft neutralization plan for interplanetary space, where xenon is used to generate the electron "fuel" internally, is based on the fact that xenon ions are charge positive and electrons are charge negative (equal positive + equal negative = spacecraft neutrality). Thus, I am going to throw the xenon ions overboard (the "other work" I mentioned a moment ago). Timed discharge of alternately positive then negative particles will work, as long as the net charge emitted over time is zero.[37]

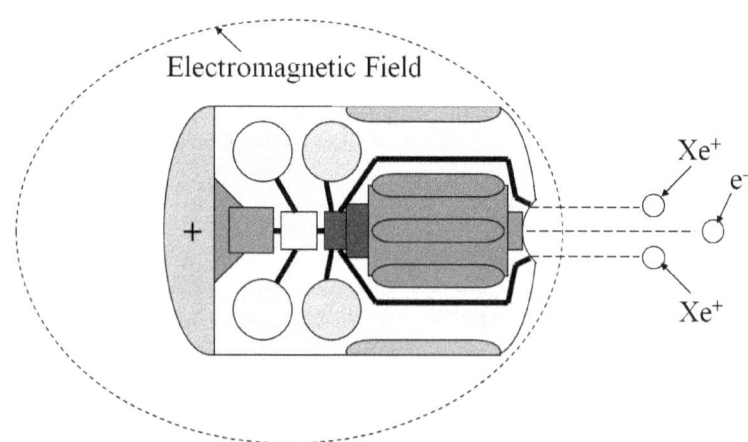

Operation in Deep Space

Primary Applications

This is a brief mention of the primary applications, to be elaborated on later. They include: spacecraft propulsion, kinetic energy asteroid/comet missile, and gravity telescope propulsion.

Crewed Spacecraft:

The performance envelope of EPT will allow it to produce a constant 1 g force (1 Earth gravity) acceleration.

$$a = \frac{qE}{m} = \frac{9.8x10^6 N}{1x10^6 kg} = 9.8 \, m/s^2$$

According to the general theory of relativity, the human body can't tell the difference between 1 g of gravity and 1 g resulting from thrust.[34] Medical problems averted by this technique are space motion sickness, heart shrinkage, calcium deficiency, muscle deterioration, coordination disruption, and space anemia.

Without the thrust generated artificial gravity advantage of EPT, there are very serious consequences. Based on rate of loss of bone density caused by microgravity (loss of 1%-3%/month), a perfectly healthy 45-year-old astronaut arriving at Mars may have the bone density of an 80-year-old. When climbing down the lander ladder, that person has a real risk of breaking their hip.[30] Houston, I've fallen, and I can't get up!

The deep space environment is a source of constant background radiation (GCR radiation). Minimizing time spent in this environment using EPT is the best way to minimize exposure. This will lower the vehicle mass requirement by reducing the amount of passive radiation shielding required for the crew module to meet the 5 rem/year radiation limitation imposed by NASA special report SP-413.[33] Based on the above arguments EPT is the absolute best option for crewed in-space travel.

Without EPT a trip to Mars without solar flares would incur 100 REMs. Among other negative effects, radiation can lower IQ, and GCR can damage the central nervous system.[30] A 2016 study found that long exposure to GCR could cause long-term cognitive dysfunction.[41] There is no practical passive shielding of GCR, however you can use water and polyethylene passive shielding for solar particle events (SPE). If shielding is too thin it can worsen exposure by increasing the number of particles. It's called spallation or shower of smaller particles. Thus, a little bit of shielding can be worse than none at all.[30]

SPEs can't be predicted (presently). If an infamous SPE like the Carrington Event of September 1, 1859 had occurred during an Apollo mission it would have killed the astronauts on the Moon.[30] The only defense against the radiation of space is a very fast spacecraft.

Given constant 1 g force acceleration, a vehicle will reach near light speed in 1 year if the acceleration remains constant.[34] This makes Electron Propulsion Technology a potential candidate for interstellar travel.

Kinetic Missile:

If Earth becomes uninhabitable by human beings, then Earth as we know it will be at an end. What to do when it's the end of all things? If a comet or asteroid of significant size hits the Earth, then it's game over! My technology is the one and only technology that can effectively neutralize the threat within the warning window allowed by modern science. One year, one EPT vehicle, no problem. The kinetic energy is more than enough to do the job. Details are in a chapter to come.

Gravity Telescope:

Finding extraterrestrial life would be the single biggest moment in human history. I have found a way to scientifically bring that moment closer. The idea of a gravitational telescope is not new, but the propulsion technology to actually make it really work is. By applying Einstein's general relativity, the Sun can be used as a lens. The minimal focal point is 550 AU (40 times the distance from the Sun to Pluto). With the capabilities of EPT it can be put into place in under three months. Today the best telescopes have a diameter of 10 m. The resulting resolution of a gravity telescope would be equal to a million miles in diameter. Sharp enough to see an alien city on an exoplanet, if there are any out there. Details are in a chapter to come.

Vehicle Concepts

The primary application for Electron Propulsion Technology (EPT) is as primary spacecraft propulsion. But, what would such a craft look like? I have come up with two models: one for uncrewed space flight and one for crewed space flight. The uncrewed EPT vehicle concept is the most developed. It seems logical to me that an uncrewed design would probably be the first to be applied in real practice. Thus, this is where I have focused my conceptualizing efforts.

Uncrewed EPT Vehicle: Conner Flyer I:

I was trying to finalize the look of my vehicle, when I saw a documentary on the 100th anniversary of the Wright Brothers' first flight at Kitty Hawk.

As such, there are some styling cues that were inspired by the Wright Brothers, even if it's not obvious. I figured that since this vehicle is essentially designed to do what no other vehicle has ever done, and I finished the visual appearance in December of 2003, the 100th anniversary month, I decided to call it the Conner Flyer I, in homage of the Wright Flyer I.

This design is unique because, having no crew to put in jeopardy; it's designed to reach EPT's highest potential velocity. Thus, this vehicle is designed to reach a relativistic velocity close to the speed-of-light. There is debate over how fast a manmade vehicle can go. This vehicle is designed to end the debate.

Isometric View

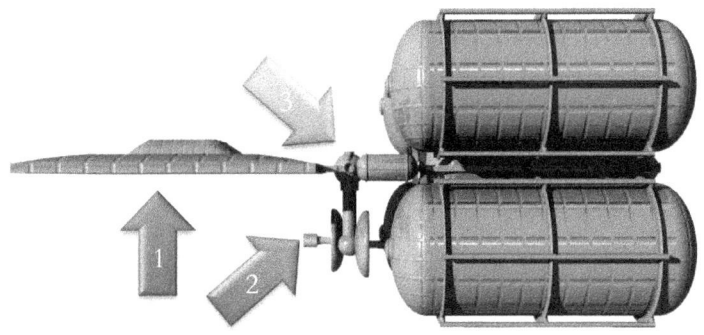

Side View

◆ 1. Power Receiving Microwave Antenna
◈ 2. Multidirectional Communication Cluster
◈ 3. Maneuvering Thrusters

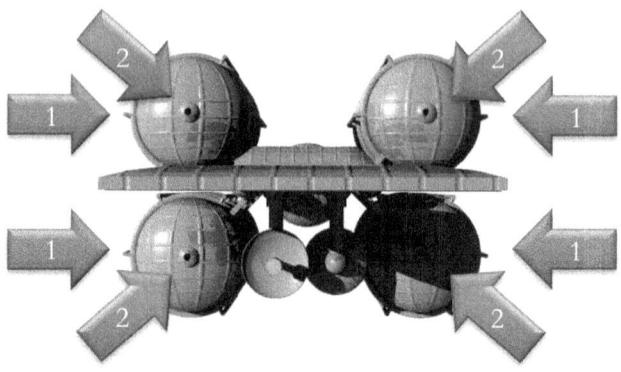

Rear View

◆ 1. Electron Propulsion Engines
◆ 2. Electron Exhaust Ports

Note: Engines in theory could be placed facing either direction. However, this orientation is preferred so that computer components and other sensitive equipment and apparatuses are most protected (by the direction of travel) from physical and radiation bombardment.

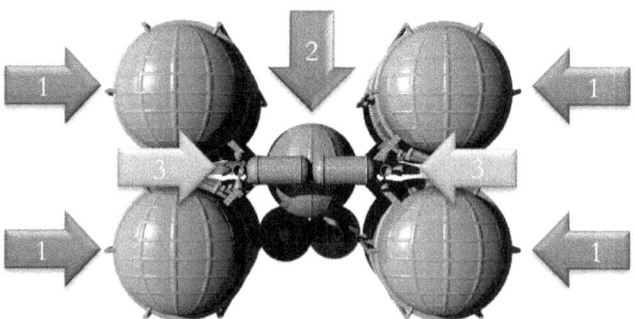

Front View

◆ 1. Electromagnetic Electron Collection Surfaces
◆ 2. Payload
◆ 3. Maneuvering Thrusters

Crewed EPT Vehicle:

The crewed vehicle is not as well thought out as the uncrewed vehicle, and I imagine it as a kind of space Winnebago (RV). Most likely a real crewed EPT vehicle would be a modification of the uncrewed EPT vehicle. However, the important function of this conceptual model is to highlight the fact that EPT has the constant thrust potential to simulate earth gravity while in route to its destination. This premise will be the cornerstone for the appearance of a crewed vehicle design.

Given that this type of vehicle can produce 1 g of thrust in both acceleration and consequential deceleration, its design must incorporate overlapping zero-g and full artificial gravity environments. When the vehicle is standing still, or at constant velocity, it will experience zero-g. However, when it's accelerating or decelerating, it will experience one full Earth gravity. It will also be in zero-g when it rotates to change direction between the acceleration phase and the deceleration phase. This will keep the artificial gravity orientation constant. Thus, while in full gravity, the floor will be the floor, and the ceiling will be the ceiling, consistently. The back wall is consistently the floor in a 1 g orientation. The real trick is designing for two gravity orientations at once.

CHAPTER 6: THRUST GRAVITY

After binge watching all the episodes of "The Expanse," on Prime Video, I've had some heavy thoughts about gravity. If not familiar, humans invent a really cool, high thrust space engine that allows for colonization into the expanse of the solar system. Once there, everything sucks, and everything goes horribly wrong. Apparently, the main form of entertainment is throwing people out of the airlock. That is until malevolent aliens open a stargate, and then comedy ensues, etc.

A key aspect of the show is humans living in different gravitational environments. In the show there are eclectic g-loads. There are natural gravitational environments like Earth (1 g), Moon (1/6 g), and Mars (1/3 g). Then there are artificial centrifugal gravity environments ($\approx 1/3$ g) like on Ceres, Eros, random asteroids, various rotating space stations, and rotating spaceships. Also, there is magnetic simulated gravity, which is basically embracing the microgravity environment (0 g), while wearing magnetic space boots. However, my favorite is thrust gravity.

As explained previously, linear acceleration can create thrust induced artificial gravity. This may sound unrealistic to maintain, but it's not that difficult. My patented propulsion system is designed specifically to induce thrust gravity.[36]

Anyway, upon thinking about gravity as a variable, I see a pattern. Full gravity (1 g) seems to be good for the body and low or zero gravity seems to be really bad. Let me unpack that a little further. Brief moments of zero gravity can produce euphoria and a generally positive experience. However, if the microgravity is prolonged negative effects occur.

Astronauts on the ISS live and work in microgravity. Some have lost as much as 20% of their bone mass. It's theorized that if humans stayed in microgravity for many generations, they could evolve to ultimately have no bones at all.[42]

Body changes can happen in just the first few days in microgravity. Things like face swelling, loss of taste and appetite, dizziness with a falling sensation, and a stuffy nose. After weeks or months, other changes happen, like a rash or eczema, possibly a fever blister, vision deterioration, maybe a cold, or allergies. This is because microgravity changes your immune system, and it can reactivate preexisting dormant viruses.[42]

In "The Expanse," lower than full gravity environments also leave their mark. Residence of Mars and the Belt (Asteroid Belt) find it difficult or impossible to return to the Earth. This is because generations of biological changes have made it essentially a one-way trip.

Ironically, adaption to the space environment could be paradoxically counterproductive to the ultimate goal of colonizing space. Earth-like exoplanets thus far tend to have a predicted Earth-like gravity. Thus, if through spacefaring the human body can't return to Earth, that also precludes colonizing Earth-like exoplanets. In other words, it's like winning the battle but losing the war.

I think gravity should be looked at differently. A little microgravity can be fun, but too much of it is really bad for your health. An analogue might be the difference between the staples of an entrée and dessert. It would be foolish to eat only dessert, all the time. Microgravity is like ice cream and cake, and full gravity is like meat and potatoes (apology to offended vegans). For maximum health, one should seek the maximum gravity (≤ 1 g). Thus, future space travel should be built around this doctrine. Simply, adapting to microgravity is a short-term thinking mistake.

CHAPTER 7: COMETS & ASTEROIDS

A technology is defined by its potential applications. The prior chapters have focused on Electron Propulsion's transportation applications. This chapter will focus on another type of application, a very special missile. This will be a brief astrophysics lesson centering on doomsday.

Armageddon

The serious possibility that a comet or asteroid could impact the Earth entered into mainstream consciousness by way of two events. The first was the televising of comet Shoemaker–Levy 9 impacting Jupiter in July of 1994. This was the first time a comet was seen impacting any planet, making the theoretical scenario very real. The second event was the 1998 blockbuster film "Armageddon." This film was likely inspired by growing concern that Earth was vulnerable to impact.

In the movie "Armageddon," a previously unknown rogue comet knocks a global killer scale asteroid into a collision course with the Earth. The world is ultimately saved by the best oil drilling team in the world landing on the asteroid using a modified next generation space shuttle. They then drill the hole and drop a giant H-bomb inside the asteroid, in the nick of time.

Certain Uncertainty

NASA has been tracking and mapping the orbits of known comets and asteroids, through projects like NEOWISE. Initially, this gave comfort to the public that the problem was all wrapped up, and there was no impending threat. Well, that didn't last very long. On February 15, 2013, at about 9:20 am, a "tiny asteroid" exploded over Chelyabinsk, Russia. The super-fireball exploded with the force of 300,000 tons of TNT! It injured over 1,000 people and damaged 3,000 buildings. There was no warning.[43]

Another alarming incident was the "Halloween Asteroid." The asteroid 2015 TB145, which was discovered on October 10, 2015, and zipped by Earth on October 31, 2015. It came within 310,000 miles of Earth, which is nearly as close as the Moon, at an unusually high speed of 78,000 mph. It was 600 m wide, which was about 30 times bigger than the asteroid that hit Chelyabinsk. Later on, NASA said it was a dead comet. The craziest part is that the object was photographed as it passed. It looked exactly like a skull.[44]

Galactic Clock

Given that asteroids can be detected and tracked, I've always considered comets the greater threat to humanity. It's true that there are periodic comets with names and known orbits. However, new comets that have never been in the inner solar system before can enter into the picture. New comets can simply just show up out of nowhere.

The anatomy of the solar system has the Sun at its center. From there is the inner solar system, which is where we live with other inner planets. Past that is the outer solar system, which contains the outer planets, asteroid belts, and a growing number of dwarf planets. All the planets that orbit the Sun do so along a disk shaped plane. All the planets are contained within a spherical structure known as the heliosphere.

The heliosphere ends at about 120 AU, which is the end of the Sun's electromagnetism.[45] On August 25, 2012, the Voyager 1 spacecraft exited the heliosphere and entered what is recognized as interstellar space.[46] However, the Sun's gravity goes out to the Oort Cloud at 20,000 AU. The Oort Cloud is a spherical structure that surrounds the solar system and contains trillions of comets.[45]

As such, the solar system should be thought of as a spherical bubble orbiting the core of the galaxy, and outer most boundary being made up of a super huge cloud of comets. This is the part where it gets weird.

The solar system is not orbiting the galaxy in the same way as planets orbit the Sun. The solar system is bouncing up and down, relative to the galactic plane. The best way to imagine this is with the metaphor of a horse on a carousel, going up and down, while the ride goes round and round. As the solar system orbits, it goes above and below the galactic star plane. The problem is that this movement can gravitationally disrupt the Oort Cloud. The result of such a disruption is a swarm of comets heading toward the inner solar system. The solar system moves up and down such that the Oort cloud could be disrupted at intervals of very roughly once every 30 million years.[47]

There is an impact history that may correlate to this pattern. About 65 million years ago was the impact incident that killed the dinosaurs (Cretaceous-Paleogene Extinction Event). It has been hypothesized that this was not an asteroid that hit the Earth in Mexico, but a comet.[48] About 35 million years ago, a comet is believed to have hit in the United States (Chesapeake Bay Impact). It's currently believed that this may be evidence of a 2 million-year-long comet shower that scientists think may have occurred between 36 and 34 million years ago.[49]

This theoretical cycle of destruction is called the Shiva Hypothesis. Shiva is the Hindu god of cyclic destruction. The theory was conceived of by NYU Prof. Michael Rampino.[47]

According to Dr. Rampino, life on Earth could be under threat, as the solar system is hurtling through a dangerous zone in the galaxy. He believes that "there is evidence that the comet activity has been high for the last one to two million years, and some comet orbits are perturbed, so we may be in a shower at the present time."[50] So basically, there is a galactic doomsday clock, that might be about to strike midnight.

Protector Missile

In considering a countermeasure, there are two main factors:

- Warning Time Issues
- How to Solve the Problem?

Thus far, alert methods for detecting threatening asteroids and comets often provide warnings that come at the last minute, or even after the fact in some cases. Luckily, none of these have been a global killer on a collision course, or else you would not be reading this now. As a result, any countermeasure will have to be quickly deployed and on continuous standby. Most proposals for defense systems to date require a minimum time scale of years to be effective. Thus, we at present are essentially defenseless against the threats we are observing.

The solution is to use Electron Propulsion Technology for missile propulsion to create a kinetic energy weapon. At conventional speeds, it can deliver conventional kinetic energy. At relativistic speeds, it can deliver relativistic kinetic energy. Such a delivery of kinetic energy to an incoming asteroid or comet can result in total vaporization.[36]

$$\text{Given: } K = \frac{m_0 c^2}{\sqrt{1 - \dfrac{v^2}{c^2}}} - m_0 c^2$$

For a vehicle of 1×10^6 kg, traveling at 0.99 c, the kinetic energy yield will be 5.5×10^{23} J. This is the equivalent of 134,623,943 megatons of TNT.

v/c	Joules/kg
0.001	4.5×10^{10}
0.005	1.1×10^{12}
0.01	4.5×10^{12}
0.02	1.8×10^{13}
0.05	1.1×10^{14}
0.10	4.5×10^{14}
0.15	1.0×10^{15}
0.20	1.8×10^{15}
0.25	3.0×10^{15}
0.30	4.3×10^{15}
0.35	6.1×10^{15}
0.40	8.2×10^{15}
0.50	1.4×10^{16}
0.60	2.3×10^{16}
0.70	3.6×10^{16}
0.80	6.0×10^{16}
0.90	1.2×10^{17}
0.95	2.0×10^{17}
0.99	5.5×10^{17}

Speed vs. Kinetic Energy

The speed at impact will be relative to the distance the object is away when it's first observed; clearly, the sooner it's observed the better. Based on Electron Propulsion Technology's incredible acceleration potential in an uncrewed configuration, it can be highly effective against a target with very little notice.

Marksmanship

The only question that remains is whether a missile can successfully be targeted at a comet or asteroid? The answer is yes, and it was answered when the NASA DART spacecraft intentionally slammed into an asteroid. The September 26, 2022 event was the first test conducted on behalf of planetary defense. The test successfully changed the trajectory of the asteroid Dimorphos.[52]

CHAPTER 8: HACKING MARS

This chapter summarizes my idea for how to hack Mars. The concept is to write an unconventional guide to terraforming Mars based on the idea that the conventional model is wrong.

I was inspired by the TV show "Mars" that aired on NatGeo in 2018. I found a few logic errors. In doing so I identified a problem requiring a solution. I created a thought experiment to solve the problem using the computer science logic that I learned in my Harvard CS50 edX course.

In my mind, I clearly identified how the terraforming model I learned as a teenager would fail epically in the face of the real facts. After deeply thinking about how it's impossible to terraform Mars, it occurred to me that a detailed analysis shows anomalies that could be exploited. Thinking like a computer hacker, and not like a scientist, per se, I realized there might be another way to come at the problem.

Dreams of Youth

I will present the model that I learned as a teenager. The idea was simple, just get in a rocket, go to Mars, and setup camp on the surface. Roasting s'mores was optional. Then over a century or more release greenhouse gases into the atmosphere until Mars is just like Earth.

Ideally, this means liquid water on the surface and a perfect polo shirt working environment. Just take your spacesuit off, and take a deep breath, and run off and enjoy the sunshine.

Listening to Elon Musk and watching the "Mars" TV show, makes it appear that they are embracing this view of terraforming. How hard could it be, right?

Heartbreaking Reality

I'll skip over the issues with getting to Mars in the first place, and just focus on the fun when you get there. Basically, it looks like Earth, but it's not like Earth at all.

At present, the commonly accepted future colonization and terraforming plans are reliant on a non-existent magnetic field. Mars' magnetosphere went dark about 4.2 billion years ago.[53] To a true Mars enthusiast, this realization is devastating.

Earth's planetary magnetic field which creates a magnetosphere is generated by a dynamo inside the planet. The rotation of the Earth produces rotating currents in the molten outer core which act like an electromagnet. Such currents can't form on Venus, because it does not rotate fast enough. The cores of Mercury and Mars are too cool. A magnetic field acts as a first line of defense against the solar wind.[54]

Without a magnetic field, the solar wind can erode away an atmosphere. However, just because a planet doesn't have a magnetic field doesn't mean that the solar wind will strip its atmosphere. Venus doesn't have a magnetic field, and despite experiencing an even stronger solar wind than Earth, it has a very thick atmosphere. Venus can hold on to its atmosphere, because it's dense and tightly bound to the planet. Mars' gravity is too weak to compress its atmosphere like that. As such its atmosphere is too thin to prevent solar wind erosion.[54]

Imagine the conventional terraforming model for a moment. Say you just released a lot a greenhouse gases to warm up the planet, and you figure out how to reconfigure the atmosphere to make it breathable. OK, so now it's exactly 74° Fahrenheit, and it's a beautiful clear day. If you take off your space helmet, your head will still explode! Well, that might be melodramatic. Just think Arnold Schwarzenegger when his helmet cracks in the movie "Total Recall."

The problems you will still have:

1. The current atmospheric pressure on Mars is now roughly 100 times less than Earth. Without a magnetosphere and given Mars' low gravity, it's unlikely that the pressure can be increased to a level whereby the air can support human and plant life.

2. The surface radiation levels will still likely be prohibitive to surface farming. Without scalable agriculture a large functioning community on Mars will be impossible.

3. Also, there is no protection from solar events. Random solar flares can damage technology and kill people on the surface.

In my opinion, the conventional terraforming model can't work on a Mars without a magnetosphere. Does that mean terraforming is impossible? Of course not. It's just that it will require more thinking and less BS.

The Hacker Mindset

The mindset of a hacker is to look for weaknesses in a system that can be exploited. The Mars problem is one that can be cracked. The magnetosphere on Mars is only "mostly" dead. It was once working. Scientists believe that Mars was once a vibrant planet with liquid water primed for life.

Mars actually does have an extremely weak magnetic field about 40 times weaker than Earth, but only in the southern hemisphere. It's coming from rocks that formed when Mars was very young and still had a magnetic field. When the planet's core cooled, and the magnetic field crapped out, it left an imprint on these old rocks. The field is too weak and nonuniform to be able to protect the planet from the solar wind.[54] However, this is an anomaly that can be exploited.

The benefit of any magnetic field will help with the very real radiation problem. The goal of a Mars colony should be to simply establish a foothold on the planet's surface. The approach will have to be focused and iterative. If the ultimate goal is terraforming, the work will have to be done in orbit, around the planet.

There are proposals for concepts like restarting the natural dynamo explosively and even an idea to literally "hot-wire" the planet.[54] The forces of nature killed the original dynamo. If you try to restart it, you are literally working against the forces of nature. I believe the logical way forward is to create an artificial magnetosphere. The simplest way to do this is with a ring apparatus.[55]

Japanese scientists believe that an artificial geomagnetic field generation can be achieved with a superconductive ring network. The model that they used was designed based on Earth. This technique was created as a possible solution to the sudden loss of the Earth's magnetosphere in the context of a polarity reversal.[55] Given that Mars has half the diameter of Earth and that superconductivity can be achieved easier with the natural cold of space, it's my opinion that applying this solution to Mars could best be done from orbit.

The artificial magnetosphere should disproportionately benefit the southern hemisphere of the planet. As such the first cities should be located here.

If you now go back to the idea of introducing those greenhouse gases, you can heat and thicken up the atmosphere. The radiation will become Earth-like. In a century, or two, or three, there will be open-air farming and walks by the lake wearing a t-shirt, jeans, and flip-flops. A little bit of mass migration later, and we will be a true two planet species.

Thinking Four Dimensionally

Now that a hacker solution for Mars has been presented, I would like to use it to enter into another thought experiment. This may be a little trippy.

Let's compare Earth and Mars as they are today. They are both part of the same solar system, so they came into existence about the same time. However, their stories are very different.

It's true that you can imagine a time before Mars lost its magnetosphere, where both planets could have been green with liquid water. Who knows? However, the biggest difference is in the two planet's moons. Mars' two moons orbit much closer to the planet and they probably were gravity captured from the asteroid belt. Earth's moon has a completely different origin story.

The Earth before the Moon is referred to as Earth Mark 1. Mark 1 had a rotation resulting in a day of roughly 18 hours as opposed to the 24 hours we have now.[56] The excepted hypothesis for the creation of the Moon is that a very large planetoid hit the Earth. The collision resulted in two new bodies: Earth Mark 2 and the Moon. The drag of the tidally locked Moon gave Earth Mark 2 the 24-hour day and its 23.5° tilt angle relative to the Sun that creates the seasons. Before the collision, Earth Mark 1 was free to change its tilt angle in ways that are not ideal for complex lifeforms.

So, the configuration of Earth today is the result of a very happy accident, which is a major factor in the existence of human life. Given this scientific understanding, isn't it weird that Mars also roughly has a 24-hour day. Actually, it's only about 37 minutes longer. Tilt is also nearly the same at a 25° angle.[57] This is generally explained away as a coincidence. Mars however doesn't have the orbital mass to result in this configuration, unlike Earth. It just spontaneously decided to copy the Earth's configuration, which resulted from a random cosmic cataclysm. Strange, right?

What if there was an explanation for Mars' planetary configuration? What if it was terraformed in the past? A ring providing an artificial magnetosphere could also be designed to change the tilt and the rotational speed. It's really practical having the two planets synchronized. What if it was a design choice?

If you don't have to answer who did it, this is a logical explanation. Basically, Mars was naturally habitable, then it lost its dynamo, then somebody came and terraformed it, then it blew up again. See, perfectly logical.

Apocalypse of the Second Genesis

Please suspend the eyerolling for a moment, as I bring up the cliché idea of ancient astronauts. There is a logical and serious argument to be made.

Humans (i.e. homo sapiens) have been on Earth for around 180,000 years.

It's believed that humans have not shown indications of neurological evolution. This means that ancient humans would have the same mental potential as anyone today. Humans have been on the Earth for a very long time. Recorded history is only roughly 5,000 years old, and the history of science is only roughly 500 years old. So, an objective and unbiased question would be, why are we not technologically farther along than we are today?

The answer in part could be something as simple as climate change. Today we are concerned about the danger of global warming. In the past there has been catastrophic global cooling. Every culture on Earth seems to have a version of the "flood myth." Perhaps the natural cycles of the Earth have destroyed an earlier version of human civilization. Then again, not unlike the concerns of today, with great technology comes the possibility of a civilization ending war.

Perhaps during the tenure of humanity, prehistory civilization reached a technological level far beyond that of our own. Given the timescales, this seems reasonable to me. So, let's imagine that spacefaring ancient people did see the importance of becoming a multiplanet species. The technological prowess required to get to Mars, design an artificial magnetosphere ring capable of altering the rotational velocity and planetary tilt, would be far in advance of where we are today. The power to create a second genesis on an alien world is no small thing.

If we go down the rabbit hole of this argument, then one has to ask what happened to the terraformed Mars? It could have been destroyed by a natural disaster, like a rogue comet, or on purpose in an act of war. Either way, if the ring was destroyed, the atmosphere would pop and blowout into space. A disaster of this magnitude would constitute an apocalyptic event for that iteration of humanity.

The Holy Grail

The magnitude of a prehistory civilization's technology that was capable of the reengineering of Mars is nothing to minimize. Imagine what that kind of technology would do for our chaotic civilization today. Currently our present-day technology can already be used to blowup the world in 20 to 30 minutes. So, there would be little downside to accessing more advanced technology, but the upside would be incredible. Think of the problems and suffering that such technology would alleviate.

The mythical quest for the Holy Grail was to directly heal King Arthur, but it was also to indirectly heal the world. Such incredible technology could actually heal our world here on Earth. So, it could be argued that there might be a Holy Grail on Mars.

X Marks the Spot

It may be reasonable to propose skipping terraforming Mars and using this new model to instead locate world saving technology.

The best way to hack Mars is to not terraform it at all, but to deductively reason out where the capital city would likely have been located. The best plan is a treasure map and a shovel.

Natural Rings

There is an interesting counterargument to my loony Mars theory. According to a new hypothesis, Mars had a little natural ring action going on at some point in the past. I pointed out that our Moon gives the Earth its tilt, which was a vital part of the development of the Human species and our civilization. Mars as well has a similar tilt but lacks the causality for such a configuration. No cause-and-effect relationship. It's just randomly so.

But now there is a theory that Mars' tilt was created by ancient rings, which later resulted in not rings but a moon. That moon was later broken up back into rings. Those rings later became moons, and so on. Such that today's two Martian moons Phobos and Deimos, are the grandchildren of the first original moon, being called proto-Phobos.[58] I guess the idea is that the ring-moon cycle orbital mass decreases over time.

This natural explanation seems plausible. It's not as fun as my thoughts on ancient astronauts or ancient aliens pursuing a terraforming project. But, minus the razzle-dazzle, it would be more logical. Definitely not as fun.

The first opportunity to explore this possibility will be when the Japanese space agency (JAXA) will acquire soil samples from Phobos in 2024. Until the samples are returned to the Earth, I will consider my crazy idea still valid.

CHAPTER 9: SURFING & EXOPLANETS

When I was learning about history in school, a running theme was the ongoing conquest to rule the "known world." This was of course with the definition of the "known world" changing as a function of time.

Conquest

So as history has gone forth, the conceptual world has literally gotten bigger. One could argue that this expansion of the conceptual world has been technologically driven. So, it's awareness that gets bigger, as the planet has always stayed the same size while humans have been on it. Maybe intelligence hasn't gotten bigger; maybe stupidity has just gotten smaller.

Travel

Air travel à la the Wright Brothers has made the idea of the world really tiny. There is nowhere in the world that is more than 24 hours away by plane. That is once you finally get out of the airport. The Internet has also contracted the idea of the world, as you can exchange thoughts with people on other continents sometimes in milliseconds.

Telescopes

Then there's our place in the stars. Since ancient times, the idea of the world has extended into the night's sky. Early on the concept of space included things like the Sun and the Moon. However, it wasn't until things took on a higher resolution through Galileo's telescope, that the physicality of other planets took hold. As the Earth was ultimately unseated as the center of the solar system, the cosmos became a defining characteristic of what the idea of the world means. As the Earth was thought to be at the center of the solar system, it was also concurrently at the center of the universe. Now the cosmos is where the Earth is located, as only a part.

In the time of Albert Einstein, the universe was only as big as our own galaxy, which was at the center of the universe. Other galaxies were thought to be funny looking spiral nebulas that where orbiting around our own galaxy. I hope you're seeing a silly symmetry by now.

Edwin Hubble proved that these curious "spiral nebulae" were in fact other galaxies. Now our galaxy was no longer alone, nor at the center of a small static universe, but just a part of an expanding cosmos.

When I was a boy, I used to look at pictures from space. They were surprisingly uninteresting. They gave the impression of vast emptiness. I always thought space should look exciting and alive. Then came the Hubble Space Telescope. When I saw the images from the telescope, after all the optic problems were solved, I thought to myself, "I knew it." The images matched how I always imagined space in my mind.

51 Pegasi

When I was born there were nine planets in our solar system. Due to downsizing and a bad economy, I think Pluto got laid off. Anyway, we still have eight. But, what's important is that there were no planets in any other solar systems. This was science fiction. That was true for all of recorded history, until 1995.

51 Pegasi b was the first recorded exoplanet found orbiting a star other than our Sun. But it wasn't the last. As of April 2016, there were 1,963 confirmed exoplanets.[59]

At present there is a very active search for exoplanets. However, if a good candidate for a world containing astrobiology is found, the opportunities for following up on that discovery are limited. Additionally, if such a candidate world were suspected of containing extraterrestrial intelligence, then the follow up options are even more limited.

As an aside, 51 Pegasi b is now no longer referred to by this name. A PR stunt by the International Astronomical Union resulted in the public renaming the planet Dimidium.[60]

WTF?

In October of 2015, it became public knowledge that the Kepler Telescope, which was designed to search for Earth-like planets, had found an anomaly that might indicate an alien civilization. The online astronomy crowdsourcing interface Planet Hunters noticed that there was something odd about the star, identified by researchers as KIC 8462852. The Planet Hunters advisory science team labeled it the "WTF star."[61] It's said to stand for "where's the flux?" I like the name because it appeals to my personal proclivity for profanity.

The strange star is roughly 1,465 light-years from Earth, or about 8.6 quadrillion miles. According to Yale postdoctoral astronomy fellow Tabetha Boyajian, "what was unusual about that was the depth of light dips, up to 20% decrease in light, and the timescales (of light variation) — a week to a couple of months."[61]

A popular explanation for the phenomenon is that it's caused by an alien megastructure. Conceptually, advanced civilizations might build planet-sized megastructures with solar panels, ring worlds, telescopes, beacons, etc. If aliens wanted to harness the energy of their home star, "they might construct enormous solar panels by the millions and send them into orbit to beam starlight down to their planet's surface." Credit for the concept goes to physicist Freeman Dyson, who imagined a giant hypothetical structure built to encompass a star. If there was an alien megastructure, "we might see the star flicker in irregular ways as the giant panels circled about it," according to Jason Wright, an assistant professor of astronomy at Penn State.[62]

SETI Results

The SETI Institute (the Search for Extraterrestrial Intelligence Institute) started investigating the star after learning that the Kepler team had vetted the "WTF star" data. They used the Allen Telescope Array, which comprises a large number of small dishes. They were searching for two different types of radio signals as possible markers for technology. The first type was narrow-band signals ("hailing signal"), and second type was broadband signals, which might indicate the presence of an alien structure in the star system. After two weeks, SETI said they had not detected any alien radio signals.[63]

Shiva Revisited

The Kepler Telescope only looked at the star in "visible light." A research team from Iowa State University looked at data from NASA's Spitzer Space Telescope, which views in infrared light. They found that the strange dips in light, which were initially theorized to be possible structures, were likely a "swarm of comets."[63]

"It's possible that a family of comets is traveling on a very long, eccentric orbit around the star," NASA said in a statement. "At the head of the pack would be a very large comet, which would have blocked the star's light."[63]

Although it might have been more fun for this to have been an alien megastructure, this observation may be more profound with a natural cause. This might be empirical evidence that the swarm of comets described in the Shiva Hypothesis is really possible. I believe, observing this phenomenon in another star system validates the premise of the theory.

Probing Pluto

The simplest way to look for suspected advanced alien life is direct viewing. If you look at the dark side of the Earth from space, it's very easy to see the city lights. However, at present high-resolution direct viewing is not possible.

One of my edX astrophysics professors suggested that the Hubble Space Telescope would be able to do this type of high-resolution direct viewing, if the light of the exoplanet's star could be effectively blocked out.[64] This would be done through an improved chronograph within the telescope, or using a space structure called a star shade. The star shade would essentially be a specialized spacecraft that would be placed in front of the telescope to strategically block the starlight.[45] So, if this worked in practice the only light would be then coming from the planet.

However, I would like to respectfully disagree with the premise of the professor's argument. Specifically, that the Hubble has the resolution to photograph a distant exoplanet. In 2002, the Hubble took a photo of the dwarf planet Pluto. It looks like a blob. In 2015, the New Horizons spacecraft took close up photos of Pluto. The Hubble photo looks ridiculous next to the new photos. Pluto is in our own solar system! Imagine how terrible the photos of an exoplanet light-years away would look. The only way to take meaningful photos of a distant exoplanet is to use a space telescope with dramatically higher resolution. In this context, my criticism also extends to the James Webb Space Telescope.

Comparison of Pluto images. Courtesy of NASA.

Voyeuristic Optics

One of the best ways to detect the presence of an extraterrestrial civilization on an exoplanet is through light signals. However, viewing the lights of an exoplanet's cities at night requires an optical telescope of 5 km in diameter.[65] Currently, the maximum size is approximately 10 m.

A solution can be found by using the Sun as a lens. As described by general relativity, gravity bends space. As such, the Sun can be used to create a gravitational lens telescope. The basic geometry of the gravitational lens of the Sun has the minimal focal length at 550 AU (astronomical units), where the light rays are brought to a focus. If a spacecraft were placed at this distance it would result in the most phenomenal telescope one can imagine. The Sun would act just like a regular lens, 1 million miles in diameter![65]

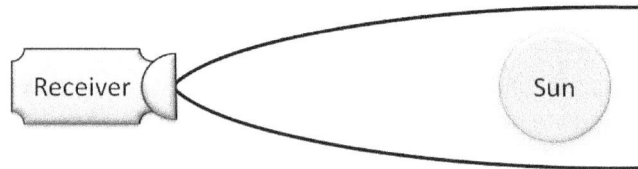

Conceptualization of Gravitational Telescope

Such a telescope would have the collecting area of 30,000 Arecibo telescopes.[65] It would be better than any radio or optical telescope you can imagine by a huge margin. The distance of 550 AU is approximately 14 times the distance to the dwarf planet Pluto. However, to get great, sharp images the spacecraft will have to be placed at 700 AU.

The resultant resolution would be high enough to detect exoplanets and photograph them in detail, and detect city lights.[65] It would also have the ability to amplify outgoing signals potentially by tens of thousands of times, theoretically allowing contact. Additional benefits would be that astrophysicists could observe highly red shifted supernovae, and continue research into the effects of dark energy.

The reason this has not been attempted in the past is that propulsion technology has been an unsolvable barrier. The farthest reaching spacecraft to date is the Voyager 1. As of February 2012, it was 120 AU from the Sun. It has been in space for about 30+ years, and is roughly 1/5 of the way to 550 AU. At its current rate of travel, it will not reach 550 AU for more than a century.

With the knowledge that Electron Propulsion Technology can produce a spacecraft that can provide a constant acceleration of 1 g, the problem is solvable.[36] The integrated application of an Electron Propulsion Engine towing a space observatory is novel, and a new concept previously unconsidered. Now all that is required is to determine the performance characteristics.

$$t = \sqrt{\frac{2x}{a}}$$

The process of sending a spacecraft to the location 550 AU requires a constant acceleration, followed by a constant deceleration. Given that 1 AU equals 1.5×10^{11} m, I multiplied this number by 550 times, minus the distance between the Earth and the Sun, which is 1 AU, for a total of 549 AU. I then divided this number by 2, to separate the acceleration and deceleration phase. After plugging this data into the appropriate formula, the resulting time from 1 g acceleration is then multiplied by 2 to give the total time required. This process was repeated for 700 AU.

$$550\ AU = 67\ days$$

$$700\ AU = 76\ days$$

Hypothetically, with given constraints the estimated time of approximately two months to two and a half months results. Thus, a gravitational lens telescope enabling spacecraft can be placed in any direction in less than three months with an Electron Propulsion Engine!

Astrobiological research would be enhanced by the use of a gravity telescope. I am not saying that it would be powerful enough to see into E.T.'s bedroom window, but it might.

OMG

The case of KIC 8462852, could have more definitively been resolved if there had been the capability of high-resolution direct detection. This is the perfect example of a situation where a gravity telescope would have been invaluable. Instead of just assuming that alien intelligence is the least possible solution, it would be possible to just look and see what is going on.

If we find evidence of extraterrestrial intelligence with our current technology, it will be tempting to explain it away, in the name of good science. However, with a gravity telescope, the observation is direct, complete, and unbiased. So with gravity telescope enabled observations, the next time a WTF star comes up, it might have to be relabeled an OMG star. As in, OMG we have a high-resolution image of an alien structure.

Super-Earths

In looking for life elsewhere in the universe, our first instinct is to look for a planetary system that is very similar to our own solar system. Then of course, just look for a planet just like the Earth. However, in reality, scientists are not finding any systems that look at all like our solar system. Furthermore, they are not finding a lot of planets just like the Earth. Actually to date, I think they have found none. What has happened in practice is that they are instead finding a lot of what are called Super-Earths. It appears that this is the most common form of hypothetically habitable planet.[45]

What makes a Super-Earth "super" is the size. Generally, they are double to triple the size of the Earth.[66] What makes a Super-Earth, or any other type of planet, habitable is that it has to be in the habitability Goldilocks zone, relative to the type of star it's orbiting. As in, "Goldilocks and the Three Bears." The planet can't be too hot, and it can't be too cold. It has to be just right for liquid water to exist on its surface.[45]

Super-Earths can come in different varieties. For example, some Super-Earths are believed to have so much liquid water, that they are actually considered "Water Worlds." There is no solid surface to the planet at all. Giant waves could travel across the entire planet, without ever breaking.[67] This type of Super-Earth would be a terrible place to live, but the surfing would be awesome!

Immortal Stars

The class of stars known as M dwarf stars, which are often red, have unique properties. When I started learning about astrophysics as a child, one of the first things I learned about was the death phases of stars. What I learned was that all stars die, but in different ways. Whether it's a supernova, the formation of a black hole, or collapsing into a neutron star, all things end. For example, the Sun, at the end of its life, will go through a death phase known as a red giant. It will grow so large, that it will consume the inner planets, including the Earth. If the Sun had significantly more mass it would go supernova, and the Earth would be blown apart. So based on what was known when I first studied, the lifespan of a planet was determined by the lifespan of its star. But, what if that star lasted forever?

M dwarfs have a lifespan of hundreds of billions of years to trillions of years.[68] This is important because it's longer than the predicted lifespan of the entire universe![69] In most cases, the universe will actually end before these stars will go through their death phase. In my mind, if something lasts as long as the universe, that is in fact the definition of forever.

So, what if you are on a planet orbiting around an immortal star? The planet is never going to be destroyed by the star! From a linear perspective, the orbiting planets inherits immortality from their parent star. However, the concept of habitability is more important. This is where it gets strange.

It appears that M dwarf stars tend to be billions of years older than our Sun.[69] So if there was life, it would have a significant head start. Given that this type of star is so prevalent in the universe, statistically if we ever discovered extraterrestrial intelligence, they would most likely be billions of years ahead of us.

It turns out that M dwarf stars may actually be more supportive of life than our star. During the earliest phases of the star's life it bombards its planets with huge amounts of high-energy radiation in UV and X-ray. To highly developed life this would be disastrous, however to life that is just starting, high-energy radiation could actually speed up evolution.[64] Later on, the UV and X-ray levels drop off way below what is emitted by our Sun. This actually creates conditions better for life than Earth. A Super-Earth around a M dwarf has even been scientifically referred to as "superhabitable."[66]

With the exception of the early phases, M dwarf stars as they are today, are both low luminosity and low UV. This allows the habitable Goldilocks zone to be much closer to the star. Thus, a planet could be orbiting incredibly close to the star, while being as habitable as Earth.[45]

Have you ever noticed that when you look at the Moon, you are always looking at "the Man in the Moon?" The same side of the Moon is always visible. The reason for this is that it's tidally locked to the Earth. Habitable planets can be so close to an M dwarf star that they can be tidally locked.

What is the life expectancy of a Goldilocks planet around an immortal star? Well, it would still be habitable today. Since it will likely not be taken out by the star, the predicted lifespan will be a function of the planet's own internal forces. If there is liquid water on the planet, hypothetically it could remain so essentially in perpetuity. However, the habitability of the planet will most likely end when the electromagnetic field of the planet collapses due to the end of its geologic activity lifecycle. It's believed that geologic activity lifecycle is a function of how large a world is. Thus, a Super-Earth would have a longevity advantage over the Earth. Such a collapse would likely be tens of billions of years from now.[68] By comparison, the Earth will be vaporized long before then.

One could speculate that if a civilization was advanced enough by the end of the planetary natural habitation period, that an artificial solution could be constructed. This could be on the planet or on a space platform of some kind. If there was an artificial means of extending habitability, then the M dwarf star would provide a comfortable environment until the end of the universe.

So, applying this knowledge, it would be a good idea to look for life around one of these M dwarf stars. What would be even better, is to find a Super-Earth in the habitable Goldilocks zone around an M dwarf star. Of course, for practical reasons, the closer to our own solar system, the better.

Wolf 1061c

On December 17, 2015, astronomers announced that they hit pay dirt. The trifecta: the right planet type, around the right star type, and right in our stellar neighborhood. Wolf 1061c exists within the habitability "Goldilocks zone." It's orbiting a red M dwarf star. But best of all, it's only 14 light-years away.[70] In cosmic terms, that is just down the street.

The Wolf 1061 star has a system of three planets. "It is a particularly exciting find because all three planets are of low enough mass to be potentially rocky and have a solid surface, and the middle planet, Wolf 1061c, sits within the "Goldilocks" zone where it might be possible for liquid water — and maybe even life — to exist," says lead study author UNSW's Dr. Duncan Wright. The mass of Wolf 1061c is 4.3 times that of Earth, the gravity is about 1.8 times that of Earth, and the solar year is about 17.9 days long. With the exoplanet's orbital distance of 0.084 AU, it's so close to its star that it's tidally locked.[71]

So, what would it be like to live on an exoplanet like Wolf 1061c? Well, it would be too cold on the dark side of the planet, so you would be living on the illuminated side. What is surreal is the fact is the sun would never set. Day would go on eternally. There would be little, if any, variations in the weather. Due to no variation in heating and cooling patterns, likely there would be no storms. You could never get a sunburn, because the UV rays are too low.[45] If you looked over a large body of water, you would notice that likely it would have no waves or tides. It would look as if the water was solid. This would be an awesome place to live, but the surfing would be terrible!

Proxima b

On August 24, 2016, it was reported that an exoplanet had been discovered by the European Southern Observatory (ESO) around our closest star, Proxima Centauri. Proxima b is located in the habitable zone of its star, such that liquid water could exist on the surface.[72] As of 2020, it's now confirmed that it's a reasonably Earth-like exoplanet, 1.17 times the mass of Earth. At a mere 4.2 light-years away, that would make it the closest potentially habitable exoplanet in the galaxy.[73]

Given that it's 20 times closer to its red M dwarf star, than the distance from the Earth to the Sun, there is a high probability that this planet is tidally locked.[73] Like the Wolf 1061c example, this configuration could be great for supporting life, but in addition to bad surfing, it would be catastrophic to our concept of the calendar.

Exoplanet Timekeeping

Our way of telling time is solely based on our situation here on the Earth. But what if we lived on a habitable exoplanet where our ideal methods of telling time don't work? Given that it's tidally locked, if you stood on the surface of the sun facing side of Proxima b. The sun would be stationary in the sky. If the sun never moves in the sky, how do you know the hour of the day? If the sun never rises or sets, how do you know what day it is?

It seems unlikely that Proxima b would have a moon, given its close distance to its sun. We use our moon to determine the month. Without a moon, how do you know what month it is?

The Earth is tilted 23.5° relative to the Sun, creating the seasons. Proxima b is likely straight up and down relative to its sun. Thus, you will never feel the change of the seasons.

Our year on Earth is determined by circumnavigating the Sun. At night, you can see the change of our position in the stars. However, if you were on Proxima b, the sunlight side sky would be perpetually opaque.

So, it would be worse than the movie "Groundhog Day." Instead of being in the same day over and over, you would be trapped in the same hour! The moral of the story is, if you're traveling to Proxima b, be sure to bring an expensive watch.

Chapter 10: Naked Wormholes

This is where advanced theoretical physics gets a little risqué. This is a concept that to my knowledge is completely novel. I baked it up in my own personal half-bakery. Though as crazy as it is, it's logical! It would open new doors of research that are worth walking through. The consequences of such a theory are more interesting than the theory!

Very Fast

Based on the special theory of relativity, it's commonly held that you can't go faster than light speed. The reason is that as your spacecraft velocity increases to nearly the speed-of-light, the mass of your vehicle will continuously increase very significantly. As such, it would take an infinite amount of energy to actually reach the exact speed-of-light.[74]

$$\gamma m_0$$

However, there may be a loophole. Not unlike a highly relativistic spacecraft, a black hole highly warps the spacetime curvature. In the Stephen Hawking book "Black Holes and Baby Universes and Other Essays," he discusses that particles can travel faster-than-light for a short distance, allowing escape from a black hole's event horizon.[75] It has occurred to me that near the event horizon that perhaps light is being bent by the black hole's massive gravity, resulting in its acceleration to a speed faster than the speed light is normally known to go. Thus, this allows, while in the black hole's spacetime influence, particles to go faster than the accepted light speed, but without going faster than the speed light is going, at the time. So, for that brief moment, they are not really superluminal in respect to the reference frame that they are traveling.

It's possible to travel faster than the speed-of-light, if light is traveling faster in the moving reference frame, than in the rest reference frame. Such that the speed-of-light hasn't been exceeded in the traveling reference frame, but has clearly been exceeded to the observer at rest.

I would describe this situation as a "wormhole," by reason that this scenario contains an interesting paradox. Both reference frames are not in violation of conventional physics; however, there is a comparative anomaly. In the spacecraft's reference frame, it never truly exceeds the speed-of-light so time is moving in a manner consistent with special relativity. However, from the observer's point of view the spacecraft disappears into a wormhole.

When the spacecraft's velocity is exactly equal to the speed-of-light, as viewed by the observer's rest frame, the value for gamma will be "no value." To calculate the relative time from the observer's point of view, it requires dividing by zero! Thus, time stops. Without the time dimension the spacecraft will no longer be observable, until its speed decreases back to below the speed-of-light for the rest frame.

$$\gamma = \frac{1}{\sqrt{1 - \frac{v^2}{c^2}}}$$

From the point of view of the observer, the spacecraft is superluminal and in a wormhole. From the point of view of the spacecraft, it's subluminal, and everything is essentially normal. So therefore, because the spacecraft is superluminal in the rest frame, but not superluminal in its own frame, there is no event horizon. The probably mythical concept of a singularity without an event horizon is called a naked singularity.[74] As such, it's logical to call a wormhole without the limitation of an event horizon, a "naked wormhole."

It's my proposition that a naked wormhole can in fact be used to calculate the total mass of the entire universe. If the speed-of-light can increase in highly curved spacetime, then there must be a non-random reason for the speed-of-light to have a constant value in the rest frame from which we conventionally observe the universe. Deductive logic would lead one to conclude that the universe must have a constant mass, which is observed through a constant speed-of-light. In other words, if the total mass of the universe were to significantly change all of the sudden, the speed-of-light would correspondingly change as well.

The experiment to prove this relationship would also empirically provide the exact total mass of the universe. A spacecraft equipped with an Electron Propulsion Engine is to be accelerated to a highly relativistic velocity. As it obtains more relativistic mass, eventually light will bend around the strong spacetime curvature it's creating. As the speed-of-light deviates upwards, it will prevent the spacecraft from becoming prohibitively massive, allowing continued acceleration. Eventually at some point, this process will result in the spacecraft reaching what is conventionally considered the speed-of-light, while remaining safely subluminal in its own reference frame.

Density Danger

From this analysis, it appears that the faster you go, the greater the effect on spacetime. This results in the acceleration of light, allowing ever increasing spacecraft velocity. This process may even be infinitely applicable to unimaginable speeds, were it not for a single factor.

As a forming black hole's gravity increases (and its corresponding spacetime curvature) its density increases radically.[38] If the theory expressed for a naked wormhole holds, then the same increase in spacecraft density should be expected. After all, its mass is increasing, while its volume remains constant, an inversion of what the forming black hole is experiencing.

It's uncertain at what point the increasing density of the spacecraft will become prohibitive and eventually fatal. I believe it will be possible to achieve speeds equal and greater than light speed (defined as c), however speeds extremely above light speed may not be possible.

Presently the speed-of-light (defined as c) is figuratively considered the speed limit of the universe. With the naked wormhole theory, density becomes the new speed limit of the universe. Hopefully someone will work around this limit as well.

Reverse Paradox

A paradox is when you get an answer that you weren't expecting. So, a reverse paradox must be really scary! One of the most famous Einstein thought experiments was the "twin paradox." A space traveling twin, having traveled at near the speed-of-light, returns to Earth to find that his sibling has aged dramatically. The "reverse paradox" turns the concept on its head. If the twin was traveling through a "naked wormhole," he might actually be older than his sibling waiting on Earth. He could be older, he could be younger, or even in a precise case, the same age! Superluminally induced rest frame temporal stagnation is to blame. Or simply put, naked wormholes are cool.

Chapter 11: Hollywood Hyperdrive

I have come up with a mechanical method to create a Hollywood-style hyperdrive system not unlike those found in the "Star Wars" and "Star Trek" movie franchises. The main characteristics of the system are hyper-acceleration, hyper-deceleration, and the ability to protect the vehicle at speed. Traditionally hyperdrives are associated with faster-than-light travel. With the combination of my "naked wormhole" theory, faster-than-light travel should be possible in conjunction with the cancelation of problematic relativistic time effects. In short, a true Hollywood hyperdrive.

Kinetic Energy

The secret to the idea is rotational kinetic energy. A rotating electromagnetic field will be placed around a spacecraft. Ideally, the inside of the field will equal zero, so that the spacetime of the vessel will be uniform. However, the outer boundary layer will experience spacetime curvature. The total energy of the spacecraft will be increased exponentially independent of linear motion. The total energy added to the system will be governed by the kinetic energy equation below.[76]

$$K = \frac{1}{2}I\omega^2$$

As the vehicle conventionally approaches the speed-of-light it experiences relativistic mass dilation. The concept is to induce this mass dilation without the vehicle moving. The rationale for this is that at relativistic speeds there are certain advantages, which are usually overshadowed by the difficulty of obtaining a relativistic speed in the first place. A powerful advantage is spacetime contraction, such that it takes less time to cover a given distance, as the relativistic distance has become smaller. Thus, given a spacecraft that can contract (or fold) spacetime while sitting still, when under way, can enjoy hyper-acceleration, while the crew feels safe levels of acceleration.

Jump Out

The result of a vehicle that hyper-accelerates is that it can literally "jump" to a speed close to the speed-of-light without hurting or killing a crew. Obviously, this process will work in reverse as well, resulting in hyper-deceleration. So, you can "jump out" of a relativistic speed as well.

Faster-Than-Light

Once a relativistic speed is achieved, it should be mathematically possible to accelerate up to a speed faster-than-light. This is discussed elsewhere in my "naked wormhole" theory, which I will not recreate here for brevity. Let it suffice to say, with this model in place, it means you can jump in and out of hyperspace, Hollywood-style. The only difference is that my math works out, and Hollywood is make-believe.

Special Perks

This system will be able to counter the negative effects of special relativity concerning time during space travel. Firstly, with hyper-acceleration and hyper-deceleration, the time of relativistic travel at a speed approaching the speed-of-light will be cut down tremendously. Secondly, in accordance with my "naked wormhole" theory, temporal stagnation will allow speeds faster-than-light travel without any time penalties. Thus, the time experienced by the traveler and stationary observer should be nearly similar. It neutralizes the negative impact of space travel predicted by the twin paradox. It should be noted that this carefully crafted plan does in fact not violate Einstein's theories or the known laws of physics.

There are additional advantages to this system. At great speeds there is the substantial hazard of colliding with small objects, mainly rocks that are too small to see, that can destroy the vehicle. The hyperdrive system as I have envisioned it will be a highly energized electromagnetic field surrounding the vehicle. Therefore, dangerous objects will either be redirected by the field distortion of spacetime, and curve around the ship, or they will be burned up by the field itself. Either way such objects will never reach the hull.

Perhaps one of the biggest mission enabling advantages, is the fact that in addition to having its own bombardment shield, it also automatically generates its own radiation shield. With such a large electromagnetic field surrounding the vehicle, it shields the vehicle from the dangers of radiation in space, from its many sources.

Tech Specs

This rotational kinetic energy hyperdrive will integrate with the already well-developed propulsion system, the Electron Propulsion Engine (EPE). This is discussed elsewhere, so I will not recreate the details of the concept here for the sake of brevity. In my opinion this is the only propulsion system concept that is adequate to use with this hyperdrive concept, based on its shear in-space output power.

To integrate the two systems, it may be necessary to transfer power between the field and the engine, within the parameters of the flight regime. I have visualized a kind of rotating swing arm within the field to power the engine, but the form of the apparatus is an inconsequential detail at this stage of development.

This hyperdrive concept is an inverse concept from the "Star Trek" warp drive, as it increases relativistic mass versus taking the total mass to zero. Thus, an EPE level system is required to break the dilated inertia.

How Much Energy?

The Electron Propulsion Engine was simultaneously designed for use with an exploration spacecraft and a comet/asteroid destroying kinetic energy missile to protect the Earth. Thus, I happen to know the exact kinetic energy output of an EPE vehicle when striking a planet-threatening object at 99% the speed-of-light. This output is described by the equation below.[38]

$$K = \frac{m_0 c^2}{\sqrt{1 - \frac{v^2}{c^2}}} - m_0 c^2$$

The resulting explosion given for a vehicle of 1×10^6 kg, traveling at 0.99 c, will be a kinetic energy yield of 5.5×10^{23} J, which is the equivalent of 134,623,943 megatons of TNT. As such, I have made the educated guess that to reach a relativistic mass dilation, while the vehicle is stationary, which is equivalent 0.99 c, the kinetic energy of rotation will have to add 5.5×10^{23} J to the spacecraft. This is not trivial.

This is not impossible, however. For reference, the kinetic energy of rotation of the Sun is 1.5×10^{36} J, the Earth is 2.5×10^{29} J, and the Crab Pulsar is 2×10^{42} J.[76] Thus, 5.5×10^{23} J for the spacecraft is less than all of these naturally occurring values.

Only Problem

Well, there is nothing that can power this system. But, hey, I shouldn't have to think of everything. The power source doesn't really matter to the theoretical validity. Space nuclear systems are good enough to run EPE, but obviously not something as power hungry as a hyperdrive. Fusion is too cumbersome for a space system of this type, as such a system would be way too massive.

In the movies, hyperdrives and warp drives run on fictional power sources. In the "Star Trek" fictional universe they use "dilithium crystals" in combination with antimatter. Obviously, dilithium crystals don't really exist and antimatter, though real, costs $62.5 trillion per gram.[65] In place of science fiction's "dilithium crystals" you need a zero-point (vacuum) energy generator. It's the only technology with the necessary power-to-mass ratio.

Future of Concept

It's an extreme, unrealistic, future technology. Technically, I believe this is possible in the next 100 years, but more likely in the next 1,000 years. What is important is that this is the correct answer, and I thought of it. For me this is a Leonardo da Vinci aerospace moment: great idea, no way in the world to pull it off. I believe it might be monetized as part of a science fiction plot structure, as it's technically unique, yet familiar in result. It's the Hollywood hyperdrive of science fiction, only it actually works!

The consequences of, when and if this system is utilizable, can't be overlooked. It's these consequences that give it Hollywood screenplay potential. In 1964 the Soviet astronomer Nikolai Kardashev developed a scale for civilizations. According to the Kardashev scale for measuring a civilizations level of technological advancement, our current civilization rates as a type 0 civilization. The American astronomer Carl Sagan more kindly placed us at type 0.7 on the Kardashev scale.[77] The successful implementation of the technology discussed in this chapter would allow us to jump straight to a type 2 civilization (Star Trek level technology). As you can imagine, that would be a weird day!

This is a novel and counter-intuitive approach to a technology that has been heretofore the sole domain of science fiction. It solves relativity generated time problems, inertia acceleration problems, and stopping problems. In concert with my "naked wormhole" theory, the result is automatically faster-than-light travel. It creates its own bombardment shield and radiation shield. The drawback is that it could take 100-1,000 years of R&D. The upside is that the result would be an automatic jump straight to a type 2 civilization, and the opportunity to leave the cosmic cradle.

THE TAO OF TOTALITY

The Falcon

In a famous scene from the 1977 movie "Star Wars: A New Hope," the character Han Solo boasts about the speed of his spacecraft, the Millennium Falcon. He says, "It's the ship that made the Kessel Run in less than 12 parsecs. ...She's fast enough for you, old man."[78] It has been noted that a unit of distance is used here, rather than a unit of time. This is generally regarded as a technical mistake of filmmaking. However, I have given some thought to this.

The Millennium Falcon uses a hyperdrive to create a spacetime wormhole, which allows it to easily zip around the galaxy. If you were to have a drag race between two wormhole creating starships, the vehicle with the best wormhole would win. Wormholes fold the space between two points. Under normal circumstances, speed is a variable, but distance is a constant. In a quarter mile race, the distance is always a quarter mile. However, in a superluminal wormhole race the distance is being contracted.

In Einstein's relativity, there are two points of view, the rest frame and the moving frame. So, to a bystander watching the race, distance is a constant, but to the racers the distance is shortening. As such victory is determined by the odometer, not the speedometer. The starship that has traveled the least distance, will cross the finish line first. As such, in context, using parsecs is the best expression of speed.

The 2018 movie "Solo: A Star Wars Story," presents a revisionist explanation. The 12 parsecs now refer to a perilous shortcut through a maelstrom. Thus, the speed of completing Kessel Run was a byproduct of a more direct route. This reinvented version is easier to understand than the original term usage. However, I believe the original relativistic interpretation is equally correct. Not to take away from "Solo," I actually liked the movie. May the force be with you.

CHAPTER 12: QUANTUM TELEPHONY

I was reading an inspiring article recently. It was about the creation of a Quantum Internet prototype being created between Stony Brook University and Brookhaven National Laboratory. It's being promoted as an unhackable network that can send information faster than the speed-of-light.[79] The principles are a little difficult to understand, but I will try to explain it based on my limited understanding.

So basically, a laser shoots out a blue high-energy photon that hits a crystal, that creates a pair of two red low-energy photons that are quantumly entangled. On a practical level, that means that if you tickle one of the pair, the other particle will giggle. Moving on. Next, one particle is then sent far away using a fiber-optic cable, while the other is kept. The retained particle is slowed down by a room temperature tube full of rubidium fog, so that the two entangled particles reach their destinations at the same time.[79]

Now a new photon is sent through a polarizer to turn it into a qubit (a quantum bit of information). The retained photon of the entangled pair and the qubit photon are both shot into a beam splitter. The result is some weird quantum three-way, whereby all the photons are entangled. The byproduct of the threesome is that the qubit data mixed with the retained photon is now magically teleported to the photon that was sent away.[79] At least that's how I understand it.

Calling Long Distance

"The basic idea of quantum entanglement is that two particles can be intimately linked to each other even if separated by billions of light-years of space; a change induced in one will affect the other."[80] So what if at a time in the future, you could make the particles you need from quantum foam in two different locations at the same time, such that they're already entangled.

Imagine if the characteristics that describe how to make an entangled particle could be written out as a long numerical sequence. This would now become your cosmic phone number. It's likely that there are multiple ways to create an entangled particle, thus there could be multiple unique phone numbers.

If you could simply "just make particles" with the amount of difficulty it takes to say microwave a burrito, then this would be mechanically realistic. The duration of the particles would not be particularly important. On demand entanglement could facilitate the ability to place a call from anywhere in the known universe to anywhere else, instantaneously.

Today's communications at near the speed-of-light allow us to send and receive information from the other side of the globe, nearly instantaneously. However, when communicating with another planet, like a rover on the surface of Mars, the limits of the speed-of-light become more obvious. In the Star Wars universe, it's routine for there to be two-way holographic communications from the core of the galaxy to say its outer rim. If you observe the speed-of-light as the cosmic speed limit, then such communications would take many, many thousands of years, one-way. On demand entanglement would allow for the creation of a "Star Wars" space phone.

PART 3: COSMOLOGY

CHAPTER 13: GRAND UNIFICATION

Though it has been a long time since I last saw Carl Sagan's "Cosmos" series, there has always been one story that stuck with me. It was the story of a man that lived long ago and measured the entire world with just a stick. At the time I found it hard to remember his name, so I just called him "stick dude."

More recently, through Google-based research, I have reconstructed the story. He was Eratosthenes (ĕr'ə-tŏs'thə-nēz') of Cyrene. And it wasn't a stick per se, but technically a staff. However, given how difficult it is for me to pronounce "Eratosthenes," I will always internally refer to him, with respect, as "stick dude."

Good Measuring

The Library of Alexandria in ancient Egypt was once considered the repository for all of the world's knowledge. Well, until it was burned down by an angry mob, of course. This is the story of one of the librarians of the Library of Alexandria, Eratosthenes of Cyrene.

Eratosthenes (276-194 BC) was a Greek mathematician, astronomer, geographer, and poet, who measured the circumference of the Earth with extraordinary accuracy. Among his teachers, was the Greek poet Callimachus in Athens.

In 240 BC, he became the head of the library at Alexandria, Egypt.[81]

Though Eratosthenes was a leading scholar in all branches of knowledge, he was considered to consistently just fall short of the highest rank in each subject. As such he was given the harsh nickname, beta.[82] This is ironic in the context of the magnitude of his historical and scientific importance today. It just goes to show, people had to take crap off their colleagues, and random stupid people, even in ancient times. I find this pointless ridicule amusing given the context of Eratosthenes' obvious greatness. Perhaps human nature doesn't change over time?

Calculation

He made a remarkably precise measurement of the size of the Earth by observing that at the summer solstice the Sun shone directly into a well at Syene (now Aswân), Egypt, at noon. At the same time, in Alexandria, Egypt, approximately 787 km due north of Syene, the angle of inclination of the Sun's rays was about 7.2°. He calculated the angle by measuring the shadow cast by a stick placed in the ground at a 90° angle. The inclination of the Sun's rays were calculated by using, angle n = arctan (length of shadow/length of stick).

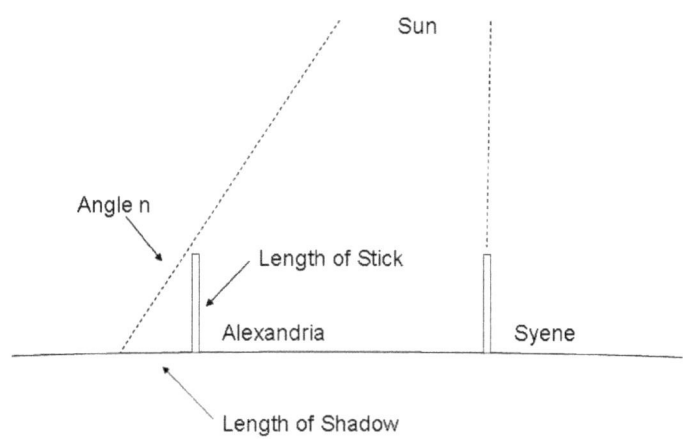

$$7.2°/360° = 787 \; km/x$$

$$x = 39{,}350 \; km$$

So, how did he do? Not bad. His answer was 39,350 km. The actual circumference is 40,076.5 km, wow![83] You have to remember, that we have the exact number now, which is roughly the exact same answer, because we live in the space age. This guy had no satellites, no GPS, no nothing. So, in layman's terms, stick plus brains equals the circumference of the Earth.

So, with this great act, mankind had once and for all proven that the world was indeed round, and would forever benefit from this knowledge. Not on your life. What I have learned from history is that it's both cyclic and sad.

Ocean Blue

In public school, I had to learn a lot of names and dates, while studying history. Ironically, Eratosthenes was not among them. The first historical fact I remember learning in school was about Christopher Columbus. I can still remember this mnemonic little charmer, "In 1492, Christopher Columbus sailed the ocean blue."

So, why did he sail the ocean blue again? Oh yeah, to prove that the world was round and to find an alternative route to India. Why did he have to "prove" that the world was round? Because, in the Middle Ages it was held as common knowledge that the world was indeed flat. So, what the hell happened? I can guess, but I don't know. It probably involved more book burning.

Pontification

I see this story as almost a parable. There should be two takeaways from this. The first is the brilliance of the human mind. How someone could take so little (a stick and an idea) and do so much. The relevant concept of taking a rudimentary tool and measuring the world will come back up in the next chapter. The second takeaway should be that the world has to be continually reminded, from age to age, that the world is not flat. This can be thought of both literally and metaphorically. Metaphorically, this might be one of those times.

I think that sometimes it pays to look back and consult ancient wisdom before plowing ahead foreword. I don't want to seem ungrateful though, I mean Christopher Columbus did discover America, and that's where I live. I guess it is what it is. However, imagine what the world would be like with a little more consistency. Technology and the Human story would probably be much farther ahead. Start, stop, then forget, seem to be the status quo.

Dark Energy & Dark Matter

What is impressive about dark energy and dark matter is that it shows how little modern science knows about cosmology. 96% of the universe is made of absolutely, completely unknown stuff! The known and observable universe only accounts for 4%, and 3.6% of that is intergalactic gas. So, all the galaxies, every star, planet, black hole, and whatever else, only accounts for a grand total of 0.4% of the universe.[84]

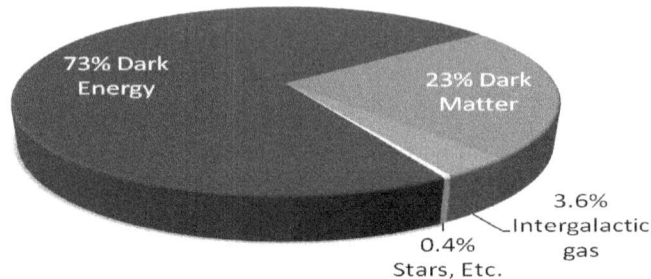

Composition of Universe

In an attempt to determine the density of the universe, a disturbing truth was learned. Two methods were used to ascertain the total density of the universe. The first was a "mass to light" ratio technique, and the second was determining mass from Kepler's laws (a.k.a. orbital analysis). The problem was that the two methods disagreed with each other, and there appears to be significantly more mass than is made apparent from the light observed.[74]

Dynamical masses (determined by orbits around galaxies and galaxy clusters) are much bigger than you would expect from the light they give off by approximately a factor of 10. This unobservable mass is referred to as "dark matter."[74]

Trying to determine the fate of the universe, two groups of researchers in 1998 discovered a disturbing mystery. There were two schools of thought on the fate of the universe. The universe began in a "Big Bang," and it could end in a "Big Crunch" or a "Big Freeze." A "Big Crunch" is where the universe recollapses into a state resembling how it would have been before the "Big Bang." A "Big Freeze" involves the universe continuing to expand without limit, until it grows cold and falls apart. It was believed that the correct scenario could be determined by measuring the rate at which the universe was slowing down. Thus, it could be determined if the universe had enough mass to come back together.[74]

"Standard candles" are required to measure distance in space. When measuring the movement of galaxies, the type Ia supernovae is the best standard candle. A type Ia supernova occurs when accreted material lands on a white dwarf from its binary partner.[74]

$$Chandrasekhar\ Limit = 1.4\ M_\odot$$

When the new material plus the mass of the star reaches the "Chandrasekhar Limit," which is 1.4 solar masses ($M_\odot$), it collapses. This type of supernovae always has the same amount of material, that all explodes at once and always has the same brightness, resulting in a "standard bomb."[74]

Using type Ia supernova standard candles, both teams learned, using different approaches, that the universe was speeding up! The conclusion was the universe is being pushed outward by "repulsive gravity," that was then coined as "dark energy."[74]

Big Rip

It's currently believed that the dark energy is constant as the universe expands. However, if the dark energy density increases as the universe gets bigger, it could result in an exponential expansion. This is a hypothetical analysis that suggests that over time the universe could essentially "explode" as the dark energy ultimately rips everything apart. This alternative scenario to the "Big Crunch" and the "Big Freeze," is known as the "Big Rip."[74]

Current Universe

The present consensus based on the concordance model of cosmology is that the total matter makes up about 25% of the universe. It's represented by the symbol Ω_M, which is the density of the total matter, both conventional and dark matter, divided by the critical density required for a "Big Crunch" turnaround.[74]

$$\Omega_M \approx .25$$

At present the current rate of cosmic expansion, also known as the Hubble constant (H_0), has been calculated to be 70 km/s/Mpc. The unit "Mpc" is megaparsecs.[74]

$$H_0 = 70 \ km/s \ /Mpc$$

Through analysis of cosmic microwave background mapping information, the following conclusion was made.[74]

$$\Omega_{Total} = \Omega_\Lambda + \Omega_M \approx 1 \pm 0.1$$

Dark Epicycles

At the birth of modern astronomy, the model of the solar system was based on a geocentric Ptolemaic view. The Earth was at the center of the universe, and the Sun and all the planets revolved around the Earth. However, the geocentric model didn't match the observations. Thus, Ptolemaic epicycles were required to make it all work out. Epicycles were smaller circles within the perfect orbital circles of the heavenly bodies. However, these modifications still didn't perfectly match the observations, so even more circles within circles were added, such that epicycles were added repeatedly.[74]

The problem of epicycles was solved by rejecting the premise that the Sun and all the planets revolved around the Earth. Copernicus introduced a heliocentric model to replace the Ptolemaic geocentric model, although this still needed some epicycles. Kepler set the planets in ellipses around the Sun, ending the process of epicycles, and providing excellent descriptive power. The explanation would later come from Newton, who derived Kepler's laws of planetary motion with Calculus.[74]

I think the moral of this story is that epicycles were necessary to defend a bad idea. By switching to a paradigm without the Earth as the center of the universe and a solar system without perfectly circular orbits, the problem was solved.

In the scientific community "epicycles" have become a swear word, referring to a theory that has become so ridiculously complex, that you just don't want to believe it anymore.[74] I would propose that the current notion of dark energy and dark matter are just modern epicycles! I also propose that they are the result of a bad cosmic model, and by just shifting the paradigm around, it can be solved. Maybe the current model is too gravity-centric?

Zero-Point Energy

In chapters 1 and 2, we looked at some of the belief systems of the Eastern world. It's logical that they would stand in stark contrast to the cutting-edge theories of our modern age. However, these wisdom traditions are ironically similar in many ways.

Here I will present the cutting-edge, forefront theory of zero-point energy. As I do so, in your mind, compare and contrast it to the fundamentals of Eastern philosophy and religions presented earlier. I think you will find the outcome amusing. I know I did.

Radio Station: Electron

As electrons orbit the nucleus of an atom, they radiate away energy, like microscopic radio antennas. According to classical physics the atom is like a miniature solar system, with electron planets and a nuclear sun. However, if this was correct then electrons would not exist, because they would spiral into the nucleus. To resolve this problem physicists had to introduce a set of mathematical rules, called quantum mechanics. Quantum theory gives matter and energy the characteristics of both a wave and a particle. It restrains the electrons to specific orbits, or energy levels, such that they can't radiate their energy unless they jump from one orbit to another.[85]

The correctness of quantum theory can be verified by the spectral lines of atoms. Atoms emit or absorb packets of light (photons), at the exact wavelength that coincides with the difference between its energy levels as predicted by quantum theory. Consequently, the majority of physicists are content simply to use quantum rules, given that they describe so accurately what happens in their experiments. However, when confronted with the question of why the electron doesn't radiate its energy away, the answer is that in quantum theory it just doesn't.[85]

The Answer is in the Void

A lot of modern physics is based on theories that work but do not answer fundamental questions. Like for example, what is gravity, why is the universe the way it is, or how did it get started? There may be answers to these seemingly unanswerable questions emerging from empty space, the vacuum, the void.

Following quantum theory, the vacuum, the space between particles of matter as well as between the stars, is not empty, as it's filled with vast amounts of fluctuating energy. "Fluctuations" is one of the most fundamental concepts to come out of the mathematics of quantum theory. This concept is the source of the uncertainty principle by Werner Heisenberg in 1927, which says that it's impossible to know everything about a system because of inherent fluctuations in the fabric of nature itself. Quantum mechanics is a statistical theory that deals with probabilities. We can't know the position and the momentum of an electron at the same time. If the momentum is known accurately, then we can determine its position only probabilistically.[85]

"Fuzziness" describes position in terms of probability waves that give a measure of the size and shape over which an electron orbit fluctuates in an atom. This also means that the energy of a particle or system is "fuzzy." Thus, there is a slight probability of it changing, or fluctuating, to another value. By fluctuation, a system can actually "tunnel" through an energy barrier, because there is a small but finite probability of the system existing on the other side of the barrier.[85]

Zero-point is an adjective that denotes that such motion exists even at a temperature of absolute zero, where no thermal agitation effects remain. Irrespective of the fact that we can't observe the zero-point energy on, say, the pendulum of a grandfather clock, because it's so minute, it's nonetheless real. This has important consequences in many physical systems.   For example, there is a certain amount of "noise" in a microwave receiver that can never be removed, no matter how perfect the technology.[85]

Zero-point energy results from the unpredictable random fluctuations of the vacuum energy, as predicted by the uncertainty principle, which is zero in classical theory.   These fluctuations can be intense enough "to cause particles to form" from the vacuum "spontaneously," provided they disappear again before violating the uncertainty principle. The formation of temporary "virtual" particles can be conceptualized like the spray that forms near a turbulent waterfall. This is termed as "quantum foam."[85]

Out of all zero-point fluctuation phenomena, zero-point fluctuations of electromagnetic energy are the easiest to detect. Electromagnetic waves have standing, or traveling modes, not unlike the modes of waves going along a rope that is shaken, where each set of waves has its own characteristic set of nodes and crests.   The zero-point energy in any particular mode of an electromagnetic field is very minute, only half a photon's worth. However, there are nearly an infinite number of possible modes of propagation, both frequencies and directions.   The zero-point energy "added up over all possible modes," therefore, is "quite enormous." It's in fact greater than the energy density of the atomic nucleus. This is in all of the so-called "empty" space around us.[85]

Despite being so large, the effects of the zero-point energy of the electromagnetic fields are not easily seen, because its density is very uniform. If a vase were standing in a true void, it would not be likely to fall over spontaneously, so as to a vase bombarded "uniformly" on all sides by packets of zero-point energy would not as well. This is because of the "balanced conditions" of the uniform bombardment. Such a barrage of energy might evidence in a minute jiggling of the vase. This is the mechanism that is thought to be involved in the quantum "jiggle" of zero-point motions.[85]

Despite the uniformity of the electromagnetic zero-point energy, there are situations where zero-point energy is slightly disturbed, and this leads to effects you can "actually measure."   An example is when the zero-point energy perturbs slightly the spectra of lines from transitions between quantum levels in atoms. This is known as the "Lamb Shift," named after the American physicist, Willis Lamb. Using techniques developed for 1940's wartime radar, he showed that the effect of zero-point fluctuations of the electromagnetic field was to jiggle the electrons slightly in their atomic orbits, resulting in a shift in frequency of transitions of about "1,000 megahertz."[85]

Another example is the "Casimir Effect." It predicts that two metal plates close together "attract each other." Imagine two plates set a certain distance apart. Between the plates, only vacuum fluctuations for which a whole number of halfwaves can exist, like waves formed by shaking a rope tied at both ends. Contrarily, outside the plates fluctuations can have many more values, because there is more space. As a consequence, the number of modes outside the plates, all of which carry energy and momentum, is greater than those inside. This imbalance "pushes the plates together."[85]

Returning back to the original question: Why an electron in a simple hydrogen atom does not radiate away as it circles the protons in its lowest-energy orbit? The electron is continually radiating away its energy as predicted by classical theory, however it's "simultaneously absorbing" a "compensating amount" of energy from the ever-present sea of zero-point energy in which the atom is immersed. The equilibrium between these two processes leads to the correct values for the parameters that define the lowest energy, or ground-state orbit. There is a "dynamic equilibrium" in which the zero-point energy stabilizes the electron in a set ground-state orbit. "It seems that the very stability of matter itself appears to depend on an underlying sea of electromagnetic zero-point energy."[85]

New Look @ Gravity

Einstein's general theory of relativity describes gravity, but we still do not know its fundamental nature. It's descriptive without revealing the underlying dynamics. This makes attempts to unify gravity with the other forces (electromagnetic, strong, and weak nuclear forces), or create a quantum theory of gravity, seemingly impossible. Attempts failed again and again on difficulties that can be traced back to a lack of understanding at a fundamental level. To overcome these difficulties theorists have resorted to ever-increasing levels of mathematical sophistication and abstraction, as in the recent development of supergravity and superstring theories.[85]

Andrei Sakharov, a well-known Soviet physicist, took a completely different direction to explain such difficulties. He suggested that rather than fundamental interaction, gravity might be a secondary or "residual" effect associated with other, non-gravitational fields. "Gravity might be an effect brought about by changes in the zero-point energy of the vacuum, due to the presence of matter." Thus, one might consider gravity as a variation on the Casimir theme, in which the pressures of background zero-point energy would be again responsible. To prove the theory, Sakharov believed that it would require predicting the value of the gravitational constant G in terms of the parameters given by zero-point energy theory.[85]

Particles sitting in the sea of electromagnetic zero-point fluctuations develop a "jitter" motion, or "Zitterbewegung," as German physicists have called it. When observing two or more particles, they are not only influenced by the fluctuating background field, but by the fields generated by the other particles as well, all similarly undergoing Zitterbewegung motion. Coupling between particles due to these fields produces the attractive gravitational force. Therefore, gravity can be understood as a sort of "long-range" Casimir force."[85]

Due to its electromagnetic underpinning, gravitational theory in this form constitutes what is known as an "already-unified" theory. This approach helps us to understand characteristics of the way gravity works that were previously unexplained. Such as: why gravity is so weak, why positive but not negative mass exists, and the fact that gravity can't be shielded because zero-point fluctuations pervade space and so can't be shielded.[85]

Wrong Answer

All though an early frontrunner, zero-point or vacuum energy has generally been excluded as a candidate for dark energy (Ω_Λ), because its total predicted energy is way too high! Most quantum field theories predict a huge value from the energy of the quantum vacuum, more than 100 orders of magnitude too large (10^{120}) to account for dark energy.[74]

Vacuum Energy $\gg \Omega_\Lambda$

At present the scientific community is completely baffled and it's anybody's guess what's going on. In the next chapter, I will present a logical and complete solution for the grand unification theory. This theory of everything is all-inclusive. Feel free to take it with as many grains of salt as you like.

CHAPTER 14: EVERYTHING THEORY

This is a discussion of the recipe for creating a good theory of everything. Like a good cocktail, there are several ingredients it must have:[86]

1. It should be able to be summarized in a single paragraph.
2. It should not use vague expressions, which can't be formulated mathematically.
3. It should build on previous theories.
4. The "theory of everything" must combine general relativity and quantum theory.
5. The idea must be testable by an experiment.
6. The "simple underlying picture" must be understandable to a layman.

Theory of Everything

I have composed a theory whereby I demonstrate that the universe itself has particle-like properties. This then is synergized with the recent completion of the Standard Model of particle physics. I have combined these two concepts together. It explains the micro to the macro, combines relativity and quantum mechanics; explaining everything in a unified field, ergo a "theory of everything."

Dark Energy & Dark Matter

I propose that dark energy and dark matter are manifestations of the same causal. It's logical to me that supermassive black holes are sources of extreme magnetism and are in fact the most powerful magnets in the universe. This has scaled symmetry as most planets (like the Earth) are magnets, and stars (like the Sun) are even larger magnets. It makes sense that as you go up the mass scale, that an object of extreme mass, would be an extreme magnet. I believe the observable results will support this concept.

As to dark matter, if it's known that stars are dipole magnets, and they were in the presents of a far more powerful dipole, would there not be an effect? Dark matter is described as "invisible glue," that holds galaxies together, but isn't conventional matter. Would it not be rational to propose that the supermassive black hole at the center of complex galaxies are magnetically bonded to the constituent stars? This would explain the higher than predicted orbital velocities of these stars.

As to dark energy, it will have the same causality. Extreme supermassive black hole magnetism can also explain the behavior of galaxy movement in the universe. A dipole has a duel effect, it attracts opposite poles, and it repels same poles. Magnetic repulsion is similar to negative gravity, which is the definition of the effect of dark energy. The proof is in the structure of galaxy formation in the universe.

Structure of Universe

The structure of the universe is basically that of soap bubble voids and dense filaments of clumps of galaxies. It's observed that these bubble/filament structures evolved over time.[74] The obvious answer to the bubbles is that they are the results of repulsion. In the early universe, there was significantly less structure, and structure became increasingly complex as a function of time. A galaxy (like a spiral) with a significant supermassive black hole at the center is a complex structure. Thus, one would assume that the effect of supermassive black hole dipoles would increase as a function of time as the universe became more complex.

Magnetic polarity has a dual effect, repulsion and attraction. So, the void bubbles are a product of galaxies pushing away from each other (same polar orientations). The filaments on the other hand are galaxies clumping as a byproduct of magnetic attraction, in combination with gravity. I would say the dominant structural force is magnetism based on the shape of the overall structure. I would say that these galaxy filaments are actually, in terms of structure, magnetic filaments. To make the point, I would cite the magnetic filaments shooting out of the surface of the Sun as the perfect analogue.

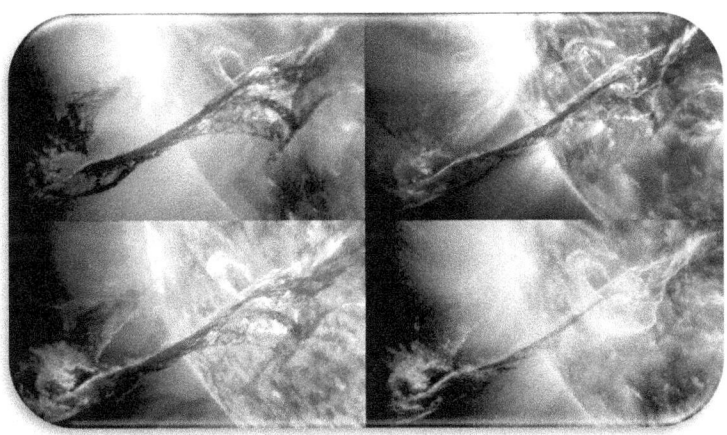

Eruption of a Giant Magnetic Filament on the Sun, 8/31/12. Images courtesy of NASA.

Movement of the Universe

The resultant movement based on this new view of the universe is as follows. It started with a "Big Bang" creation moment, which is extrapolated to an initial singularity. This initial singularity starts with a scaling factor of essentially zero.[74]

Big Bang:

$$0 + \varepsilon$$

The early universe was significantly denser and hotter than present. It's predicted that it was at a temperature so hot that hydrogen ionized, making the very early universe opaque. This early stage of the universe has been observed as the cosmic microwave background (CMB). Observations are impossible earlier than a redshift of $z = 3,000$.[74]

It's conventionally believed that, when the universe was smaller the dark matter was greater than the dark energy, given the assumption the dark energy is a constant. Thus, it's believed that the density of dark energy and dark matter changes with time and behaves differently as you change the size of the universe. So, in the early observable past the universe would initially be decelerating.[74]

Based on my hypothesis that dark energy is the repulsion of same polarity galaxy cores, you would expect this influence to be minimal in a less complexly structured universe. In a more simplistic cosmos gravity is the default influence.

Beyond CMB:

$$\Omega_\Lambda < \Omega_M$$

Presently the universe is accelerating, so it's known that at some moment in the past the density of dark energy and dark matter were balanced. Turnaround point is observed at a redshift of z = 0.8.[74]

$$\Omega_\Lambda = \Omega_M$$

In the present moment the universe is expanding and accelerating, thus the density of dark energy is greater than the density of dark matter.[74]

With structural complexity being a function of time, supermassive black hole magnetism is now the principal universe structuring force. The galaxies are pushing away from other galaxies, while simultaneously clustering with other galaxies, resulting in the magnetic bubbles and filaments shapes.

Present:

$$\Omega_\Lambda > \Omega_M$$

In the future of the cosmos, with polarity attraction and repulsion having positioned the galaxies in a fashion where neighboring galaxies are oriented such that there are not repelling one another, the more familiar effect of gravity should regain its dominance. As proximity and orientation based repulsion slows, there should be another turnaround point where the density of dark energy and dark matter are balanced.

Future:

$$\Omega_\Lambda = \Omega_M$$

If you go with the idea that dark energy is supermassive black hole magnetic repulsion, then a logical conclusion is that there is nothing to prevent galaxies, which have been non-uniformly placed together in filament structures from combining together. As this galaxy clumping takes place the universe will go from complex to simple, allowing gravity to be the dominant force.

$$\Omega_\Lambda < \Omega_M$$

The resultant conclusion based on this new view of the universe is a "Big Crunch," which is extrapolated to a final singularity. This final singularity ends with a scaling factor of essentially zero.[74]

Big Crunch:

$$0 + \varepsilon$$

An example of the concept of the universe converging in a gravity dominated way after a period of dark energy dominated expansion is our own galaxy the Milky Way. The Milky Way galaxy is moving closer to its celestial neighbor the Andromeda galaxy. "Very interestingly, we find that the Andromeda galaxy does appear to be coming straight at us," said Roeland van der Marel, an astronomer at the Space Telescope Science Institute in Baltimore. The discovery was made thanks to images taken over the lifespan of the Hubble Space Telescope.[87]

NASA scientists say they know "with certainty" when our beloved galaxy will cease to exist as we know it. The new data from the Hubble Space Telescope proves, NASA says, that in 4 billion years the Milky Way and Andromeda will collide or pass each other by so closely that the gravitational force each exerts on the other will cause them to slow down to the point of merging. The merger will be completed 6 billion years from now. It's the massive gravitational pull that ultimately drew the Milky Way and Andromeda together, and will ultimately cause them to become one. "The clear finding is, we are going to merge with Andromeda," van der Marel said. "In the past, it was just a possibility, but now it is a known fact that this will happen."[87]

Harmonic Oscillator

To understand the implications of this new view of the universe, a graphical representation is helpful.

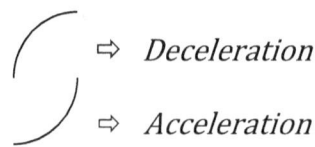

$\Rightarrow$ *Deceleration*

$\Rightarrow$ *Acceleration*

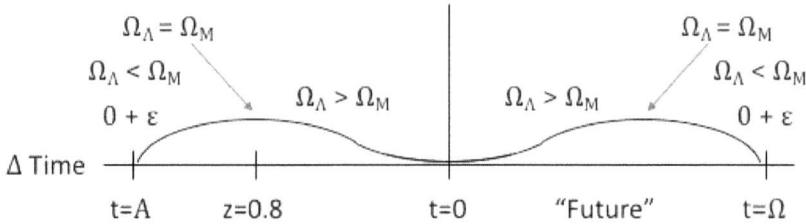

Δ Acceleration

$\Omega_\Lambda = \Omega_M$ $\Omega_\Lambda = \Omega_M$

$\Omega_\Lambda < \Omega_M$ $\Omega_\Lambda < \Omega_M$

$0 + \varepsilon$ $\Omega_\Lambda > \Omega_M$ $\Omega_\Lambda > \Omega_M$ $0 + \varepsilon$

Δ Time

     t=A   z=0.8   t=0   "Future"   t=Ω

Sinusoidal Movement of the Universe: "Alpha (A) to Omega (Ω)"

Proportional deceleration, acceleration and deceleration describes a sinusoidal wave function. As such I have concluded that the universe is a harmonic oscillator. This movement is quantum mechanical.[88]

Test #1: Gravitational Telescope

To define the curvature of the movement of the universe accurately, a better space telescope is required. Such a telescope must exceed the current generation of space telescopes. A solution can be found by using the Sun as a lens (as explained in chapter 9). As described by general relativity, gravity bends space. As such, the Sun can be used to create a gravitational lens telescope. The basic geometry of the gravitational lens of the Sun has the minimal focal length at 550 AU (astronomical units), where the light rays are brought to a focus. If a spacecraft were placed at this distance it would result in the most phenomenal telescope one can imagine. The Sun would act just like a regular lens, 1 million miles in diameter![65]

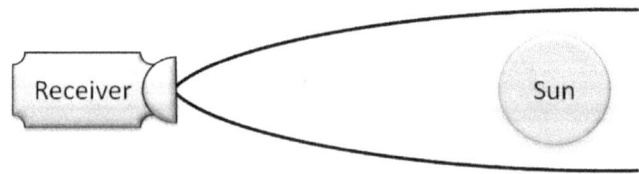

Conceptualization of Gravitational Telescope

Hypothetically, in combination with my patented propulsion technology the estimated application time of approximately two months to two and a half months results. Thus, a gravitational lens telescope enabling spacecraft can be placed in any direction in less than three months with an Electron Propulsion Engine!

$$550\ AU = 67\ days$$

$$700\ AU = 76\ days$$

So, if you are looking for high redshift supernovae you don't care where you point, they're everywhere. You just point at some nice black piece of space, take really deep pictures of it, and wait for the supernovae to start showing up.[74]

Parameters of Science

Science is the study of the observable and testable universe. An "event horizon" constitutes an edge to the universe. As such the scientific (non-philosophical) universe extends from the cosmic event horizon inwards, with exclusions of spacetime contained within black hole event horizons. Thus, the universe is a kind of Swiss cheese of finite, definable spacetime.[74]

It's logical that if the universe's spacetime is finite and definable, that there must be a finite total mass value. I don't mean a proportional mass ratio, but a real value that can be empirically obtained, and then applied as a computational constant.

Test #2: Naked Wormhole

It's my proposition that a naked wormhole (as explained in chapter 10) can in fact be used to calculate the total mass of the entire universe. If the speed-of-light can increase in highly curved spacetime, then there must be a non-random reason for the speed-of-light to have a constant value in the rest frame from which we conventionally observe the universe. Deductive logic would lead one to conclude that the universe must have a constant mass, which is observed through a constant speed-of-light. In other words, if the total mass of the universe were to significantly change all of the sudden, the speed-of-light would correspondingly change as well.

The experiment to prove this relationship would also empirically provide the exact total mass of the universe. A spacecraft equipped with an Electron Propulsion Engine is to be accelerated to a highly relativistic velocity. As it obtains more relativistic mass, eventually light will bend around the strong spacetime curvature it's creating. As the speed-of-light deviates upwards, it will prevent the spacecraft from becoming prohibitively massive, allowing continued acceleration. Eventually at some point, this process will result in the spacecraft reaching what is conventionally considered the speed-of-light, while remaining safely subluminal in its own reference frame.

Electron Propulsion Engine Performance:

The Electron Propulsion Engine (EPE) was designed from inception to reach a high relativistic velocity. Relativistic effects are defined by γ, between two inertial reference frames of differing velocities. However, while the reference frame system is accelerating this effect is more intimidatingly represented by γ^3. The following is the equation for EPE traveling at 0.99 c at 1 g.

$$a = \frac{qE}{\gamma^3 m_0} = \frac{(358\ C)(9.8x10^6\ N/C)}{(7.1)^3(1x10^6 kg)}$$

$$= 9.8\ m/s^2$$

Notably at 99% the speed-of-light, EPE is still at 1 g. As such, it's possible to use EPE to reach the highest possible percent of light speed for empirical testing of the spacetime medium. The test would end when the total acceleration becomes negligible, or contrary to convention a superluminal speed relative to the rest frame is achieved.

Eratosthenes Nouveau:

It's stick dude all over again. You may recall from the previous chapter that Eratosthenes was the ancient mathematician who measured the circumference of the Earth accurately over 2,000 years ago with a stick. Conceptually this is not entirely dissimilar to the experiment of Eratosthenes. You have the symmetry of measuring the very large with the very small. An elegant idea executed with a rudimentary tool. My "stick" is my Electron Propulsion Technology spacecraft. However, instead of measuring the world, I'm going to measure the whole cosmos. This is a vintage way of thinking that is wise to repeat.

Test Results:

The telemetry data collected from this event will prove the constant total mass of the universe. If you know the amount of relativistic mass change in the spacecraft, and the corresponding, proportional change in the speed light is traveling in the spacecraft's reference frame (ŕ), then you can calculatively deduce the total mass of the universe (M_U), resulting in a proportional rest constant for the recognized speed-of-light (r).

$$\Delta m_0 \; \alpha \; \Delta c_0, \, in \; \acute{r}$$

$$\therefore \; M_U \; \alpha \; c, \, in \; r$$

Relevance of Total Mass

Knowing the total mass of the universe is important, because if the speed-of-light is defined by the total mass, this has implications elsewhere in physics. Specifically, the Planck relation demonstrates that the Planck constant would be different if the speed-of-light was not the conventional value.

$$E = \frac{hc}{\lambda}$$

Thus, it can be learned why a quantum is a given value from the point of view of its causality. The twelve recognized elementary particles are specific quanta based values. All of which are conditional on the speed-of-light. Thus, the causality of the non-random elementary particles of quantum mechanics can be reconciled, quite ironically, by an extreme relativity experiment. This theory is moving closer to unity.

Defining the Universe

The Standard Model of particle physics mediates the dynamics of known subatomic particles, concerning the electromagnetic, weak, and strong nuclear interactions. The current formulation was finalized in the mid-1970s upon the discovery of quarks. The Standard Model has been strengthened by further discoveries: bottom quark in 1977, top quark in 1995 and the tau neutrino in 2000. With its success in explaining experimental results, it's regarded as a "theory of almost everything." The Standard Model can't be considered a complete theory of everything, because it doesn't incorporate dark energy, gravitation as described by general relativity, and dark matter.[89]

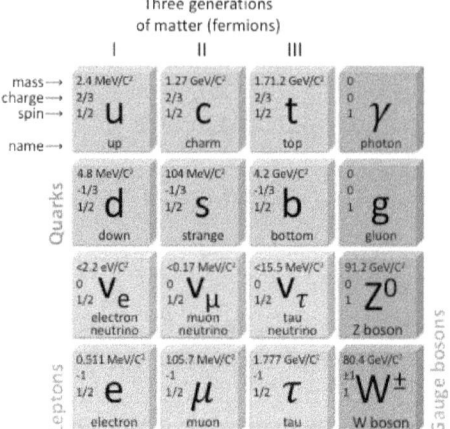

Three generations
of matter (fermions)

Standard Model of Elementary Particles

The Standard Model can be thought of as the "Newton's laws of particle physics," if you will. It's composed of 12 particles of matter stuck together by 4 forces of nature. The resulting equation can explain why the sky is blue, why atomic nuclei stick together, and why DNA is the shape it is.[90] The equation is as follows:

$$\mathcal{L}_{Standard\ Model} =$$

$$-\tfrac{1}{2}\partial_\nu g_\mu^a \partial_\nu g_\mu^a - g_s f^{abc} \partial_\mu g_\nu^a g_\mu^b g_\nu^c - \tfrac{1}{4} g_s^2 f^{abc} f^{ade} g_\mu^b g_\nu^c g_\mu^d g_\nu^e +$$

$$\tfrac{1}{2} i g_s^2 \left(\bar{q}_i^\sigma \gamma^\mu q_j^\sigma\right) g_\mu^a + \bar{G}^a \partial^2 G^a + g_s f^{abc} \partial_\mu \bar{G}^a G^b g_\mu^c - \partial_\nu W_\mu^+ \partial_\nu W_\mu^- -$$

$$M^2 W_\mu^+ W_\mu^- - \tfrac{1}{2}\partial_\nu Z_\mu^0 \partial_\nu Z_\mu^0 - \tfrac{1}{2c_w^2} M^2 Z_\mu^0 Z_\mu^0 - \tfrac{1}{2}\partial_\mu A_\nu \partial_\mu A_\nu - \tfrac{1}{2}\partial_\mu H \partial_\mu H -$$

$$\tfrac{1}{2} M_h^2 H^2 - \partial_\mu \phi^+ \partial_\mu \phi^- - M^2 \phi^+ \phi^- - \tfrac{1}{2}\partial_\mu \phi^0 \partial_\mu \phi^0 - \tfrac{1}{2c_w^2} M \phi^0 \phi^0 - \beta_h \left[\tfrac{2M^2}{g^2} +\right.$$

$$\tfrac{2M}{g} H + \tfrac{1}{2}(H^2 + \phi^0 \phi^0 + 2\phi^+ \phi^-)] + \tfrac{2M^4}{g^2}\alpha_h - ig c_w [\partial_\nu Z_\mu^0 (W_\mu^+ W_\nu^- -$$

$$W_\nu^+ W_\mu^-) - Z_\nu^0 (W_\mu^+ \partial_\nu W_\mu^- - W_\mu^- \partial_\nu W_\mu^+) + Z_\mu^0 (W_\nu^+ \partial_\nu W_\mu^- -$$

$$W_\nu^- \partial_\nu W_\mu^+)] - ig s_w [\partial_\nu A_\mu (W_\mu^+ W_\nu^- - W_\nu^+ W_\mu^-) - A_\nu (W_\mu^+ \partial_\nu W_\mu^- -$$

$$W_\mu^- \partial_\nu W_\mu^+) + A_\mu (W_\nu^+ \partial_\nu W_\mu^- - W_\nu^- \partial_\nu W_\mu^+)] - \tfrac{1}{2} g^2 W_\mu^+ W_\mu^- W_\nu^+ W_\nu^- +$$

$$\tfrac{1}{2} g^2 W_\mu^+ W_\nu^- W_\mu^+ W_\nu^- + g^2 c_w^2 (Z_\mu^0 W_\mu^+ Z_\nu^0 W_\nu^- - Z_\mu^0 Z_\mu^0 W_\nu^+ W_\nu^-) +$$

$$g^2 s_w^2 (A_\mu W_\mu^+ A_\nu W_\nu^- - A_\mu A_\mu W_\nu^+ W_\nu^-) + g^2 s_w c_w [A_\mu Z_\nu^0 (W_\mu^+ W_\nu^- -$$

$$W_\nu^+ W_\mu^-) - 2 A_\mu Z_\mu^0 W_\nu^+ W_\nu^-] - g\alpha [H^3 + H\phi^0 \phi^0 + 2H\phi^+ \phi^-] -$$

$$\tfrac{1}{8} g^2 \alpha_h [H^4 + (\phi^0)^4 + 4(\phi^+ \phi^-)^2 + 4(\phi^0)^2 \phi^+ \phi^- + 4H^2 \phi^+ \phi^- + 2(\phi^0)^2 H^2] -$$

$$gMW_\mu^+ W_\mu^- H - \tfrac{1}{2} g \tfrac{M}{c_w^2} Z_\mu^0 Z_\mu^0 H - \tfrac{1}{2} ig [W_\mu^+ (\phi^0 \partial_\mu \phi^- - \phi^- \partial_\mu \phi^0) -$$

$$W_\mu^- (\phi^0 \partial_\mu \phi^+ - \phi^+ \partial_\mu \phi^0)] + \tfrac{1}{2} g [W_\mu^+ (H\partial_\mu \phi^- - \phi^- \partial_\mu H) - W_\mu^- (H\partial_\mu \phi^+ -$$

$$\phi^+\partial_\mu H)] + \tfrac{1}{2}g\tfrac{1}{c_w}(Z_\mu^0(H\partial_\mu\phi^0 - \phi^0\partial_\mu H) - ig\tfrac{s_w^2}{c_w}MZ_\mu^0(W_\mu^+\phi^- - W_\mu^-\phi^+) +$$

$$igs_wMA_\mu(W_\mu^+\phi^- - W_\mu^-\phi^+) - ig\tfrac{1-2c_w^2}{2c_w}Z_\mu^0(\phi^+\partial_\mu\phi^- - \phi^-\partial_\mu\phi^+) +$$

$$igs_wA_\mu(\phi^+\partial_\mu\phi^- - \phi^-\partial_\mu\phi^+) - \tfrac{1}{4}g^2W_\mu^+W_\mu^-[H^2 + (\phi^0)^2 + 2\phi^+\phi^-] -$$

$$\tfrac{1}{4}g^2\tfrac{1}{c_w^2}Z_\mu^0Z_\mu^0[H^2 + (\phi^0)^2 + 2(2s_w^2 - 1)^2\phi^+\phi^-] - \tfrac{1}{2}g^2\tfrac{s_w^2}{c_w}Z_\mu^0\phi^0(W_\mu^+\phi^- +$$

$$W_\mu^-\phi^+) - \tfrac{1}{2}ig^2\tfrac{s_w^2}{c_w}Z_\mu^0H(W_\mu^+\phi^- - W_\mu^-\phi^+) + \tfrac{1}{2}g^2s_wA_\mu\phi^0(W_\mu^+\phi^- +$$

$$W_\mu^-\phi^+) + \tfrac{1}{2}ig^2s_wA_\mu H(W_\mu^+\phi^- - W_\mu^-\phi^+) - g^2\tfrac{s_w}{c_w}(2c_w^2 - 1)Z_\mu^0A_\mu\phi^+\phi^- -$$

$$g^1s_w^2A_\mu A_\mu\phi^+\phi^- - \bar{e}^\lambda(\gamma\partial + m_e^\lambda)e^\lambda - \bar{v}^\lambda\gamma\partial v^\lambda - \bar{u}_j^\lambda(\gamma\partial + m_u^\lambda)u_j^\lambda - \bar{d}_j^\lambda(\gamma\partial +$$

$$m_d^\lambda)d_j^\lambda + igs_wA_\mu\left[-(\bar{e}^\lambda\gamma e^\lambda) + \tfrac{2}{3}(\bar{u}_j^\lambda\gamma u_j^\lambda) - \tfrac{1}{3}(\bar{d}_j^\lambda\gamma d_j^\lambda)\right] + \tfrac{ig}{4c_w}Z_\mu^0[(\bar{v}^\lambda\gamma^\mu(1 +$$

$$\gamma^5)v^\lambda) + (\bar{e}^\lambda\gamma^\mu(4s_w^2 - 1 - \gamma^5)e^\lambda) + \left(\bar{u}_j^\lambda\gamma^\mu\left(\tfrac{4}{3}s_w^2 - 1 - \gamma^5\right)u_j^\lambda\right) +$$

$$(\bar{d}_j^\lambda\gamma^\mu(1 - \tfrac{8}{3}s_w^2 - \gamma^5)d_j^\lambda)] + \tfrac{ig}{2\sqrt{2}}W_\mu^+[(\bar{v}^\lambda\gamma^\mu(1 + \gamma^5)e^\lambda) + (\bar{u}_j^\lambda\gamma^\mu(1 +$$

$$\gamma^5)C_{\lambda K}d_j^K)] + \tfrac{ig}{2\sqrt{2}}W_\mu^-\left[(\bar{e}^\lambda\gamma^\mu(1 + \gamma^5)v^\lambda) + (\bar{d}_j^K C_{\lambda K}^\dagger\gamma^\mu(1 + \gamma^5)u_j^\lambda)\right] +$$

$$\tfrac{ig}{2\sqrt{2}}\tfrac{m_e^\lambda}{M}\left[-\phi^+(\bar{v}^\lambda(1 - \gamma^5)e^\lambda) + \phi^-(\bar{e}^\lambda(1 + \gamma^5)v^\lambda)\right] - \tfrac{g}{2}\tfrac{m_e^\lambda}{M}[H(\bar{e}^\lambda e^\lambda) +$$

$$i\phi^0(\bar{e}^\lambda\gamma^5 e^\lambda)] + \tfrac{ig}{2M\sqrt{2}}\phi^+[-m_d^K(\bar{u}_j^\lambda C_{\lambda K}(1 - \gamma^5)d_j^K) + m_u^\lambda(\bar{u}_j^\lambda C_{\lambda K}(1 +$$

$$\gamma^5)d_j^K] + \tfrac{ig}{2M\sqrt{2}}\phi^-[m_d^\lambda(\bar{d}_j^\lambda C_{\lambda K}^\dagger(1 + \gamma^5)u_j^K) - m_u^K(\bar{d}_j^\lambda C_{\lambda K}^\dagger(1 - \gamma^5)u_j^K] -$$

$$\tfrac{g}{2}\tfrac{m_u^\lambda}{M}H(\bar{u}_j^\lambda u_j^\lambda) - \tfrac{g}{2}\tfrac{m_d^\lambda}{M}H(\bar{d}_j^\lambda d_j^\lambda) + \tfrac{ig}{2}\tfrac{m_u^\lambda}{M}\phi^0(\bar{u}_j^\lambda\gamma^5 u_j^\lambda) - \tfrac{ig}{2}\tfrac{m_d^\lambda}{M}\phi^0(\bar{d}_j^\lambda\gamma^5 d_j^\lambda) +$$

$$\bar{X}^+(\partial^2 - M^2)X^+ + \bar{X}^-(\partial^2 - M^2)X^- + \bar{X}^0\left(\partial^2 - \tfrac{M^2}{c_w^2}\right)X^0 + \bar{Y}\partial^2 Y +$$

$$igc_wW_\mu^+(\partial_\mu\bar{X}^0X^- - \partial_\mu\bar{X}^+X^0) + igs_wW_\mu^+(\partial_\mu\bar{Y}X^- - \partial_\mu\bar{X}^+Y) +$$

$$igc_wW_\mu^+(\partial_\mu\bar{X}^-X^0 - \partial_\mu\bar{X}^0X^+) + igs_wW_\mu^-(\partial_\mu\bar{X}^-Y - \partial_\mu\bar{Y}X^+) +$$

$$igc_wZ_\mu^0(\partial_\mu\bar{X}^+X^+ - \partial_\mu\bar{X}^-X^-) + igs_wA_\mu(\partial_\mu\bar{X}^+X^+ - \partial_\mu\bar{X}^-X^-) -$$

$$\tfrac{1}{2}gM\left[\bar{X}^+X^+H + \bar{X}^-X^-H + \tfrac{1}{c_w^2}\bar{X}^0X^0H\right] + \tfrac{1-2c_w^2}{2c_w}igM[\bar{X}^+X^0\phi^+ -$$

$$\bar{X}^-X^0\phi^-] + \tfrac{1}{2c_w}igM[\bar{X}^0X^-\phi^+ - \bar{X}^0X^+\phi^-] + igMs_w[\bar{X}^0X^-\phi^+ -$$

$$\bar{X}^0X^+\phi^-] + \tfrac{1}{2}igM[\bar{X}^+X^+\phi^0 - \bar{X}^-X^-\phi^0]$$

So, this equation seems pretty complete. Hey, wait a minute. There is something off. When you see the "H" in the equation, this is the Higgs particle. As of the writing of this equation, the Higgs had yet to be discovered. No Higgs equals a big problem.

The Higgs particle is a hypothetical massive scalar elementary particle and is a key building block in the Standard Model. It has no intrinsic spin, and for that reason is classified as a boson.[89]

The Higgs boson plays a unique role in the Standard Model, by explaining why the other elementary particles, except the photon and gluon, are massive. In electroweak theory, the Higgs boson generates the masses of the leptons (electron, muon, and tau) and quarks. It's believed that fundamental particles are massive because they are surrounded by Higgs particles, resulting in the Higgs field.[90]

Large Hadron Collider

In order to investigate the unresolved mysteries of particle physics, the scientific community has constructed the most amazing technological device in history, the Large Hadron Collider (LHC). In Switzerland, filling the Geneva countryside, it's hidden 100 m down, in a tunnel 27 km in circumference. It's one of the coolest places in the entire galaxy (-271°C), one of the hottest places in the universe (10 million billion°C), and has a vacuum emptier than outer space. It's the most technologically advanced machine ever created, with the largest number of high-tech magnets ever built, and it's the largest, most complex, electronic instrument on the planet.[91] The LHC is the most powerful particle accelerator in the world, with 9,000 magnets. Every second 2 billion protons smash into each other.[91] So, what is it looking for?

String theory theorists are hoping that the LHC's proton head-on-collisions will prove their theory. If there is enough energy, it may eject some debris into extra dimensions predicted by string theory. Measuring the amount of energy after the collision, and comparing it to the amount before, it can be determined if there is less energy after than before. If there is less, it will be evidence that the energy has drifted away.[92]

Other theorists are looking for supersymmetric particles to explain dark matter. Additionally, it's thought the LHC might actually produce dark matter particles. Because they are conceptualized as non-interactive, they will escape the detector, so their signature will be missing energy.[84] Yes, the "missing energy" thing again.

The LHC is the biggest scientific experiment ever attempted. The most amazing fact about the LHC is what it has not found! It has found no supersymmetry, no dark matter, and no secret dimensions.[91] However, the primary goal of the LHC is to look for the elusive Higgs particle. It's predicted by and is necessary for the Standard Model of particle physics.[90]

Higgs Particle Discovery

As it's said, "seek and ye shall find." On July 4, 2012, I had the opportunity to watch the live webcast of the official announcement of the discovery of the Higgs particle on my iPad. CERN announced the discovery of a "Higgs-like boson." They noted that it will take years for confirmation, with the LHC set for "restart" in 2015. This is the LHC's first and only major discovery to date.

Although this completes the Standard Model, it's not the ultimate particle theory, as you need more than the Standard Model to complete a "theory of everything." There is nothing to explain why the Standard Model works the way it does, as the so called "God particle" imbues substance (gives state), as the Higgs generates masses. Additionally, the Standard Model doesn't say why particles have certain masses.

During the Q&A session, it was pointed out that the Standard Model alone was incomplete, and didn't take into account something like the graviton particle. A particle like the graviton can't be detected by a particle collider.

In the apparent absence of proof of other cosmological models, if the graviton were somehow found, then it would allow the unification of general relativity and the Standard Model, literally resulting in a complete "theory of everything." What to do, what to do?

Universe is a Graviton

To be straight forward, to take the completed Standard Model from where it is now to a complete explanation of the universe, one particle is missing. I have postulated that the cosmos as a whole is behaving in a particle-like fashion. I believe I have found the missing particle. The universe itself is a cosmic-scale graviton boson. Like a graviton, a cosmic-scale particle would have no independent mass separate from the total mass of the universe (mass = 0). Like a graviton, a cosmic-scale particle would have no charge, as the universe is charge neutral and the summation of all charges (charge = 0). It's reasonable to assume that the spin of the universe is an integer spin, and bosons have no antiparticle. To the best of my knowledge there is nothing that would prohibit this unorthodox interpretation.

The consequence of the universe being a graviton is unity between general relativity and the Standard Model within a single framework of quantum field theory. The ultimate test for any "theory of everything" is whether it can be expressed in an equation no longer than an inch. See below:

$$E = \frac{hc_{M_U}}{\lambda}$$

Conclusion

Einstein's Maxim:

"How glorious it is to discover an underlying unity in a set of phenomena that seem completely separate."[93]

There are a lot of ideas in this chapter, so I will recap by pulling them all together. The Standard Model of particle physics was presented, and it was demonstrated that it's almost a complete theory of everything, were it not for specific factors. These factors include that it fails to address dark energy, dark matter, and gravity as defined by general relativity.

I started by presenting the idea that in fact dark energy and dark matter are the result of a single phenomenon. When the implications were analyzed, it was demonstrated that the universe is moving quantum mechanically like a particle. I went further to define the characteristics of this cosmic scale particle. I demonstrated how to utilize currently patented technology to more accurately (than the current scientific data) define the exact movement of the universe. I continued to explain how to utilize this same patented technology to calculate the exact mass of the universe.

A cosmic scale graviton particle is deduced from an analysis of dark energy and dark matter. The resulting movement is derived based on the interplay between the cosmological constant and the force of gravity as defined by general relativity. As such the cosmic scale graviton particle automatically incorporates dark energy, gravitation as described by general relativity, and dark matter. Therefore, when the cosmic scale graviton particle concept is added to the Standard Model, it's automatically a complete theory of everything.

In fundamental physics "beauty" is a very successful criterion for choosing the right theory.[94] It's simple, and thus it has elegance. It has beauty because it doesn't incorporate or require exotic complications. There are no strings, no secret dimensions, or any other strange things. Reality defined in regard to three spatial dimensions and one time dimension.

An important attribute of my "theory of everything" is that it encompasses previous theories of what is already known. This is specifically in regard to Einstein's general relativity and the Standard Model of particle physics. It in no way says previous science is wrong. Additionally, it vindicates the cosmological constant, and is 100% compatible with the Standard Model of elementary particles.

My theory explains everything from why the sky is blue, to why a quanta is a quanta, to why the speed-of-light is what it is. The universe, one field, unified.

Golden Paragraph

As I promised at the beginning of the chapter, I can summarize this theory of everything into one coherent paragraph:

The Standard Model of particle physics is almost a complete theory of everything, were it not for specific factors. These factors include that it fails to address dark energy, dark matter, and gravity as defined by general relativity. Extreme supermassive black hole magnetism can explain the behavior of galaxy movement over time in the universe, which is observed as dark energy, and supermassive black holes at the center of complex galaxies are also magnetically bonded to the constituent stars, resulting in the observation of dark matter. Resulting movement of the universe from this analysis results in the observation that the universe as a whole is acting like a harmonic oscillator. This movement is quantum mechanical. I propose that the universe is itself acting like a particle of physics with the properties of a graviton. The consequence of the universe being a graviton is unity between general relativity and the Standard Model within a single framework of quantum field theory. As such the speed-of-light value of "c" is defined relative to the total mass of the universe. This theory can be tested by using Electron Propulsion Engine technology to calculate the exact movement of the universe, by way of a gravity telescope. Additionally, Electron Propulsion can be used to measure the exact mass of the universe through superluminal performance data collected while the vehicle is in a naked wormhole. Therefore, when the "quantified" cosmic-scale graviton particle concept is added to the Standard Model, it's automatically a complete theory of everything.

CHAPTER 15: AFTERGLOW

This chapter represents further thought, illuminated by the light of my complete "theory of everything." The beauty of a theory of everything is that it allows you to look at the universe holistically. As such I have developed a kind of meta-universal dynamics perspective.

Zero-Point Power

The ultimate secret of zero-point energy is the concept of resonance. The basic concept is that current (power) is a function of frequency. In a circuit when you are off-resonance the current is very low. If a circuit is modified to be on-resonance, the current can suddenly increase dramatically. The "quality" of resonance can narrow the resonance frequency, such that the current can suddenly increase in a potentially destructive manner. The resonance frequency is expressed in the below equation.[95]

$$\omega_0 = \frac{1}{\sqrt{LC}}$$

Resonance frequencies are tuned in by radios and televisions by a variable capacitor. In this way, power can be "tuned in." Thus, we don't have access to zero-point vacuum energy today, simply because we are "off-resonance!"

Off-Resonance

I have no idea how to make a zero-point energy generator, mechanically, however there may be some hints from prehistory. I personally have a secular/Zen perspective on life, however out of boredom and a sense of impending death I decided to read the "King James Bible" cover-to-cover. I figured it couldn't hurt. In the "Book of Jude" in the "New Testament" of the "King James Bible" it makes reference to another book, "The Book of Enoch the Prophet." In my opinion, in this context the book is presented as the "code crib" to decipher the "Book of Revelations." Unfortunately, it was lost to history until the beginning of the 17th century, with the first English translation in 1821. As such it can be speculated that most interpretations of the "Book of Revelations" have been made without the code key. But I digress.

In "The Book of Enoch the Prophet," he talks about the Archon Angels (Watchers) that teach man the "secret of the crystals." For this crime they are referred to as the Satans. They are said to be eternally damned for the joint crime of teaching the knowledge of Heaven to the primitive humans, and having sex with the Earth women. The result of the latter crime resulted in their offspring becoming Nephilim giants. The matter was ultimately resolved by Noah's flood, which killed all the giants and erased the knowledge of the crystals.[96] Not saying I believe any of that; I am just throwing it out there.

To continue the theme Plato of ancient Greek fame also had a similar story. In the dialogues of "Timaeus" and "Critias," Plato describes the legendary island of Atlantis. To my knowledge, Plato does not specifically refer to Atlantis as being high-tech. However, Edgar Cayce in a 1923 reading made mention of an advanced Atlantis civilization that was powered by mysterious form of energy crystal. According to Plato Atlantis was destroyed by being submerged under the sea. This is not particularly different from being destroyed by a flood.

If I dig further into prehistory accounts, I am sure that I will find more of the story pattern of a high-tech past involving crystals that is destroyed by water. The point is that if you are trying to "tune in" an energy resonance, a crystal is a logical choice.

So, theoretically it might be possible to use crystals, of some type, to hit quantum resonance, which results in an exponential energy spike powered by zero-point vacuum energies. Think of it as a radio that gives you infinite energy, by tapping into energy that is already there.

Possible Results

The initial status upgrade from the use of this technology would go from a type 0 civilization straight to a type 2 civilization (Star Trek level technology) on the Kardashev scale. However, this development is relative to the efficiency of production and miniaturization. Theoretically, I believe with a higher level of miniaturization and some time, it could potentially result in a type 3 civilization (Star Wars level technology).

Spiritual Component

This part of the chapter actually scares me, as it truly pushes the envelope. Entering into resonance in the context of zero-point energy release should, in principle, be accessible both mechanically and organically. Organic beings (people) should be able to theoretically realign with the universe, such that they are in resonance. In doing so, they would become a different type of being. The result of such an occurrence would result in a type 4 civilization (Bible story level technology).

Diamond in the M.U.D.

While preparing to write this chapter, it amused me to realize that the initials for the concept of "meta-universal dynamics" spells "mud." This is amusing because I read a Buddhist story once about a guy that was a drunkard found by a Buddhist monk. It was called "Finding a Diamond on a Muddy Road."

Zen master Gudo of Kyoto, the spiritual teacher of the emperor of Japan, was walking on his way to Edo. There was a heavy rain and he stopped in a village to buy some new sandals to replace his wet ones. Instead he was given the sandals for free and was taken in for the night by a housewife of a family that appeared to be depressed. The housewife said that her husband was a drunkard and a gambler.[97]

The Zen master gave the housewife some money and asked her to get a gallon of wine and something to eat. When the husband returned around midnight yelling for his wife to make dinner, he was greeted by the monk who gave him wine and fish. The man drank all the wine and passed out on the floor. In the morning the monk said, "Everything in life is impermanent. Life is very brief. If you keep on gambling and drinking, you will have no time to accomplish anything else, and you will cause your family to suffer too." The husband responded, "you are right."[97]

The perception of the husband awoke as if from a dream. The husband offered to escort the monk for a short distance and carry his things. The man followed Gudo for the rest of his life. All modern Zen teachers in Japan spring from the lineage of the successor to Gudo. They called him Mu-nan, which means, the man who never turned back.[97]

It's a Buddhist imperative to get out of the mud of the world. Of course, first you have to understand the mud. The concept of meta-universal dynamics explains the nature of the metaphorical mud, as proper application of the concept would result in escaping the nature of the human condition, literally. It represents both the cause and solution to how we are experiencing our reality. There is a way out of the mud!

Beyond Science

When I was a very little boy, I started out with a fundamental question that no one could answer. The question was constructed like this: "I live on a planet called the Earth that circles the Sun, and the Sun is in a galaxy called the Milky Way, and our galaxy is one of many galaxies, and the galaxies are in the universe." Okay, this is fair enough. "What is outside the universe?"

Outside Universe

This is a seemingly innocent question, but the implications are complex. I would say that if you can't define where the universe is, then you can logically argue that the universe, relative to its external environment, is nowhere. If the universe is nowhere, then the galaxies are nowhere, and the Milky Way is nowhere, and the Sun is nowhere, and the Earth is nowhere. If the Earth is nowhere, then my house is nowhere, and then that means I am nowhere. If I can be defined as existing nowhere, then it's logical that I don't exist.

When I was five years old, I became interested in astrophysics as a direct result of "Star Wars." However, my previously stated question was my prime driver to study cosmology specifically. What has disturbed me is that my parents couldn't answer the question satisfactorily, my teachers at public school couldn't answer the question, my professors at university couldn't answer the question, and the correct answer is not in any book I have ever read. My childhood perspective was, how can I live my life if it can't be rationally argued that I exist. Perhaps I was a little "mental" as a child, but I could feel myself ungrounded floating in indefinable, non-localized space. I couldn't believe no one else was disturbed by this "obvious" problem. Eventually, I told my brain to shut up, and focused on pursuing the stated goals of life. The caveat was that I promised myself that one day before I died, I would try to figure out this issue.

In my "theory of everything," I present the universe as having a finite mass. This finite mass can be empirically determined, resulting in a cosmological mass constant. I have been able to forward the idea that the causality for a constant mass, and thus other subsequent constants, is the unifying force of motion. I also have been able to address dark energy and dark matter, and make a logical argument as to the ultimate fate of the universe. However, despite this, I am alarmed to realize that I still have not answered my original question that started my scientific quest in the first place.

What I have not answered is what is moving, what is outside the universe, and thus where is the universe? The answer of modern science is that there is a "cosmic event horizon" where the universe ends. Thus, it's unanswerable by science as to what is outside the cosmic event horizon, as science can only exist inside the universe. This then leaves such a question as purview of religion or philosophy. However, I haven't been presented with an answer from these disciplines either.

Collectively the only answer I have received from the world is a combination of "I don't know," "stop asking," and "who cares." The following is an attempt to go a little farther than the conventional perspective.

Delta Assemblage

My "theory of everything," which I believe is conceptually complete as it's presented, is essentially an ultimate measurement of the universe as a whole. However, any measurement is only as good as its degree of uncertainty! The greatest potential source of error is the observer. Perhaps it's the correct answer from the point of view of all human observers, however if a more expansive viewpoint was possible, then my result would be incomplete in regard to ultimate totality.

The issue is quantum mechanics, as the observer in essence "assembles" reality from a number of possibilities. As such in an assembled reality, how do you know that your version of reality contains all the information to comprehensively put together a complete "theory of everything?"

Given that all of the Latin and Greek letters have been exhausted, I decided that it would be easier to use a Sanskrit symbol. As such I have used the symbol "Om" (ॐ), which can be loosely translated as the vibrating manifestation of absolute reality. Delta (Δ) is the Greek symbol used to symbolize change or the possibility of change. Thus, the concept that the universe in total could change relative to the state of the observer is stated as "delta assemblage" and symbolized by Δॐ; solely for the purposes of this book. Thus, our agreed upon assemblage is symbolized as follows:

$$\text{ॐ}_0$$

As such, the degree of uncertainty for the measurements called for in my "theory of everything," is as follows:

Theory of Everything $\pm \Delta$ॐ$_0$

In my "theory of everything," I demonstrate that the universe is a particle. Particles must be assembled by an observer. From this there are four primary consequences:

1. Different Assemblage Possible
2. Lower Assemblage Possible

3. Higher Assemblage Possible
4. No Assemblage Possible

In the books of the bestselling author Carlos Castaneda, he describes the Native American Toltec perspective that reality can be disassembled and reassembled differently. Specifically, that "other worlds," what science fiction would call parallel dimensions, can be assembled. This perspective is not in violation of quantum mechanics. So, one could ask the question: Is my "theory of everything" still true in all alternate parallel assemblage realities? I would hope the answer is yes.

$$\Delta \, \text{ॐ} \approx \text{ॐ}_0$$

Further in the books of Carlos Castaneda, his mentor known by the pseudonym Don Juan Matus, states that lower states of assemblages of reality are possible. The ancient Toltecs believed that a human could briefly experience reality from that point of view of a lower animal state. From this one could philosophically entertain the notion that reality could be assembled at lower energy levels. Such a perception when applied to cosmology might result in a less complete perception of the universe.

$$\Delta \, \text{ॐ} < \text{ॐ}_0$$

To extend the thought experiment even farther, one could imagine that the universe could be assembled at higher energy levels than humans currently do. Would this reveal otherwise hidden intrinsic properties? Such a perception when applied to cosmology might result in a more complete perception of the universe. As such this could affect my "theory of everything." In other words, there might be more "everything" to a higher life form. Don Juan would call such things "unknowable."

$$\Delta \, \text{ॐ} > \text{ॐ}_0$$

Quantum mechanics teaches that particles can also exist as waves. Thus, if the universe is a particle, could it be perceptually disassembled, and observed as a wave? In other words, viewed as pure energy with no form.

$$\Delta \, \text{ॐ} \gg \text{ॐ}_0$$

Thus, the above thought experiments give rise to a fifth consequence:

5. Reality is as Real as You Make It

My "theory of everything" can be the key that unlocks the door to the ultimate secrets. However, it's from a single (common) point of assemblage, thus it can't be total understanding. Ultimate truth is thus impeded by current human limitation.

$$\Delta \overset{\circ}{\mathfrak{F}}_0 = \Delta \text{Universe}$$

The Buddha Siddhartha stated this concept more concisely in the "Supreme Mantra" contained at the end of the "Heart Sutra."

གཏེ་གཏེ་པྲ་ར་གཏེ་པྲ་རྞ་སཾ་གཏེ་བོ་དྷི་སྭ་ཧཱ།

ga-te ga-te para-gate para-samgate bodhi SVAHA!

"Gone, gone, gone beyond, gone altogether beyond!
O enlightenment! Be it so! Hail!"[98]

- The Buddha, from the Heart Sutra

The Veil

My theory is mathematically and scientifically correct, yet it feels like there is something missing. It feels like there is still some subtle thing left to discover that will make this intellectual adventure complete.

To satisfy my personal curiosity I read the "The Nag Hammadi Library." It's the text that inspired the "Matrix" movies. It can be thought of as the Bible Part 2. My only complaint was that it seemed incomplete. I think there should have been a Bible Part 3. But I digress. In this text it introduces a concept that could actually answer my original question!

In "On the Origin of the World," it states that our perceptible universe is the "Abyss of Nothingness." It's also referred to as the "Chaos of the Abyss of Darkness." It's a pocket of empty darkness, contained within the substance of the totality, referred to in "The Gospel of the Egyptians" as the "Light of Everything." The realm of light is called "The Pleroma," which translates to fullness, which obviously is the opposite of emptiness. The realm of light and darkness are said to be separated by "the veil."[99]

What if there was something between our galaxy and the rest of the universe. Would it not skew and limit our perception of the universe as a whole? It's possible if our assumptions are based on a darkness-centric based error, that it could result in a flawed cosmic model? Perhaps the reason I have not found a satisfying answer to my original question, was because the model of the universe I was presented with as a child was exceptionally incomplete. How will this affect my theory? The best way to sort it all out is to further test the logic of my theory of everything.

Surprise Ending

While studying at edX, I took an astrophysics course. It was mostly for fun. It was conducted by Australian National University. Having completed my theory of everything a year earlier, I thought it might be interesting to ask some probing questions. As such, I started having a brief email chat with one of the course professors. To keep the discussion simple, I only mentioned dark matter. I wrote the following:

Dark Matter Question:

"If the black hole at the center of the galaxy were a crazy large electromagnet, it would explain dark matter. This assumes that the field extends out beyond the stars of the galaxy. The intense electromagnetism can curve spacetime with a gravitational result. The interior of such a field would be of the same strength within the sphere of the field. I believe that this hypothesis is consistent with Maxwell's laws. Anyway, just a random thought. Not sure if this is possible?"

Professor's Reply:

"I guess this would be true, but it would require the magnetic field to be so strong that its energy density would exceed the density of matter. Such a field would be very easy to measure (we can measure the magnetic field in space in a variety of ways). In fact, it would probably need to be so strong it would warp compasses on Earth!"[100]

My Reply:

"Thank you for answering my question. I was very happy with your thoughtful response. I knew my idea was a long shot. However, allow me to close with a few whimsical thoughts.
The source of the electromagnetism is in fact infinitely dense, it's a black hole after all. If our galaxy were producing a very strong electromagnetic field it could potentially block out or shield the readings of the fields of other galaxies. You wouldn't know you were in a very strong galactic field because it's uniform. So, my compass needle would still point north."

Professor's Reply:

"Nice idea, but the very nature of a magnetic field is that it has a direction - there is no way to have such a field and not notice. Of course, there could be some other sort of hitherto unknown field that does the trick... "[100]

My Reply:

"Okay, this is going to be far-fetched. What if you had the galactic central black hole creating a very strong electromagnetic field that created a natural Faraday cage around the star plane. The electromagnetism from other galaxies doing the same thing would be blocked, while at the same time slowly varying magnetic fields (like that of the Earth) would not be interfered with. This isn't a sane thought, but it's a valiant last try. This argument does, however, seem to match up with the observations of dark matter."

Not surprisingly I never got a reply back. However, that didn't deter me from completing my thought.

Of course, I mentioned in my online conversation that all galaxies would be doing the same thing. It's generally considered unscientific to imagine that a unique situation could exist. We must live on an unexceptional planet, around an unexceptional star, in an unexceptional galaxy. Thus, as to our galaxy, then as to all galaxies. However, this is an assumption based on no scientific data.

If in fact this was an oddball galaxy, we would never know it. If in fact there was a Faraday cage around only our galaxy, then this would be actual scientific proof of "the veil" from the "The Nag Hammadi Library." But again, I digress.

My Follow-Up:

"I know you don't agree with my reasoning. But here is a little physical evidence to support my concept. [YouTube link] It just popped up. There's no need to reply.

In trying to debunk my own theory, I figured to have a Faraday cage you need to have a conductor wrapped around the galaxy's star plane. But, if it was something obvious like dust and gas it would shine. Well ironically, that appears to be the new discovery."

The important takeaway from this is that in order to reconcile my theory of everything, I had to make a prediction. Out of pure luck, I had a YouTube subscription to Science@NASA, and the proof that my prediction was right simply showed up out of nowhere.

It's Glowing

The 2014 "ScienceCast" was called "The Milky Way is Not Just a Refrigerator Magnet." The NASA video explained that researchers working with data from the European Space Agency's Planck spacecraft had mapped the magnetic field of the entire galaxy. "The Milky Way's magnetic field stretches across more than a hundred thousand light-years," says Charles Lawrence, the U.S. Planck project scientist at NASA's Jet Propulsion Laboratory."[101]

In late 2013, Planck wrapped up a mission to study the cosmic microwave background radiation, the afterglow of the Big Bang. However, to study the cosmic microwave background, Planck had to unravel everything in the foreground, and that includes microwave emissions from the Milky Way.[101]

It went on to say that interstellar dust of the Milky Way "shines with a microwave light that is polarized by galactic magnetism." "The Planck map, which resembles a giant fingerprint, is proof of a galactic dynamo at work."[101]

I can't believe it. The galaxy is glowing. Victory!

CHAPTER 16: THEORY DEFENSE

This chapter is a defense of my "theory of everything." In the space of time since writing my theory, observations have been published that further support the predictions of my theory. One day the preponderance of material hit "critical mass," such that I was compelled to write down my concluding thoughts on the matter. There are three principled arguments that can be made against my theory, pursuant to my conclusions about dark energy and dark matter.

1. That the Milky Way is not an oddball galaxy.

2. That my theory's reliance on galactic and intergalactic magnetism is far-fetched.

3. That the current understanding of black holes is complete and beyond any criticism.

I will attempt to satisfy these concerns, by providing evidence that our home galaxy is a weirdo, that a cosmic web a magnetism is slowly being revealed, and the that the modern conceptualization of black holes is antiquated.

Uniqueness Spectrum

I have hypothesized through my "theory of everything" that our galaxy is an oddball in the cosmos. Perhaps it would be helpful to think of galactic uniqueness as a spectrum, with typically boring on one end and five-alarm oddball on the other. Perhaps with specialness somewhere in between.

Warped & Twisted

Normally we think of galaxies as being as "flat as a pancake." Our neighbor galaxy Andromeda appears to be flat. However, the Milky Way isn't, it's "warped and twisted."

Astronomers from Macquarie University and the Chinese Academy of Sciences found that the Milky Way gets "increasingly warped and twisted the further away the stars are from the galaxy's center."[102]

Astronomers don't want to think of our home galaxy as in any way "special," however, its twisted shape gives it a "specialness." It's duly noted that, "our Milky Way's twists are rare, but not unobserved elsewhere in the universe."[102]

So, just how warped and twisted is it? The Earth is located roughly 27,000 light-years from the galactic core. The warp "starts at ranges about 25,000 light-years from the galactic core, and it gets more severe with distance." Stars at 60,000 light-years from the galactic core "are as far as 5,000 light-years above or below the galactic plane." Our galaxy "gets thicker with distance." Where we are in the galaxy it's only 500 light-years thick, however at the outer edges it's "as much as 3,000 light-years thick."[103]

Disproportionate Darkness

A research team using data from ESA's Gaia Mission and observations from the Hubble Space Telescope has calculated the Milky Way's galactic mass by applying Kepler's laws of orbital motion to the observed outer-galaxy orbital velocity. They concluded that the figure is about 1.5 trillion solar masses. However, there are only 200 billion stars in the galaxy. Throwing in the mass of the galactic core's supermassive black hole only adds another 4 million solar masses. After you add in a bunch of dust and gas it still results in 90% of the galaxy being dark matter, which is nuts.[104]

Here is a reality check. Our galaxy is roughly 129,000 light-years across, and our neighbor galaxy Andromeda is roughly 220,000.[104,105] We have roughly 200 billion stars, and Andromeda has approximately 1 trillion. Yet, it's estimated that Andromeda is only 800 billion solar masses.[104] Which obviously is far less than our smaller galaxy.

In roughly 4 billion years, the Milky Way and Andromeda will collide. It has long been assumed that this would be the end of the Milky Way. However, if the data is to be believed, Andromeda will be "subsumed into the Milky Way."[104] Given that the size and mass of our home galaxy don't match, I conclude that it's officially bizarre and unexpected. I believe we are in oddball territory now.

Cosmic Web

Magnetism has been observed by the Max Planck Institute for Radio Astronomy in galaxies and between galaxies. Magnetic forces in galaxies determine "whether stars cluster in spiral arms or are grouped in an elliptical fashion." Observations "support the idea that galaxy magnetic fields are generated by a dynamo process."[106]

For the first time magnetic fields have been seen between two galactic clusters, which "suggests some of the largest scale structures in the universe are magnetized." About 1 billion light-years away a field between galaxy clusters Abell 0399 and Abell 0401 was observed by the Low-Frequency Array radio telescope network (LOFAR), based mainly in the Netherlands. Radiation from the "electrons zipping through the magnetic fields revealed this magnetism." The magnetism is inside a gaseous filament that connects the clusters in a "cosmic web."[107]

Giant magnetic fields in the universe have been observed with magnetic structures several million light-years in extent. The "emission originates in an extremely ordered magnetic field."[108] Magnetic fields "exist in the center regions of a galaxy, where a supermassive black hole resides in nearly all cases."[109]

Astrophysicists at Western University believe that supermassive black holes actually predate stars as the first structures after the Big Bang. Stellar-mass black holes are formed by stars that weigh greater than 20 times the Sun going supernova. At the center of galaxies are supermassive black holes that weigh millions, and in some cases billions, of times that of the Sun. It had been believed that supermassive black holes grew slowly over time, however new observations suggest "that these giants were already in place and massive long before the first stars ever formed."[110]

The explanation is that the supermassive black holes formed from "direct-collapse black holes." The black holes would have collapsed from very large clouds of gas. Some of these supermassive black holes were "already the mass of billions of Suns just 800 million years after the Big Bang."[110]

Given that direct-collapse supermassive black holes predate all other structures in the universe, and are responsible for galactic formation, is it really so far-fetched that supermassive black holes could be powerful magnetic dynamos?

Steam Engines

Initially theoretical physicists in the early 1970s, like Stephen Hawking and Jacob Bekenstein saw thermodynamics as a useful analogy for understanding black holes, but later claimed that "it's an identity." This was the beginning of black hole thermodynamics. They basically grabbed a thermodynamics textbook, took the laws, and replaced thermodynamic terms with black hole variables. "It says that black hole laws, most of which are features of the geometry of space-time, are somehow identical to the physical principles underlying the physics of steam engines."[111]

I believe that black holes, especially supermassive black holes, are far more interesting and important than 19th century steam engines. This seems to be an unnecessarily limiting belief based on a runaway analogy.

Conclusion

Yes, I'm right that the Milky Way is an oddball. I'm also right that in the context of galactic and intergalactic magnetism, supermassive black holes are the most relevant objects to the overall structure of the universe. Additionally, I'm right that black holes are more dynamic and interesting than 19th century steam engines. It's good to be right.

CHAPTER 17: TOTALITY

Like in the movie/novel "The Hitchhiker's Guide to the Galaxy," if the question about the secret of the universe is insufficient, then the answer will be also lacking. As such, I have not only found an answer to the theory of everything, as traditionally formulated, but I have taken it one step further. By improving the question, there is a more sophisticated and satisfying solution.

Maligning the Multiverse

When asked about what is outside the universe, a common answer, at present, is the multiverse. The universe by definition is everything, and outside of everything are independent everythings. If you ask a scientist how many universes are in the multiverse, they will say infinity. This cleverly avoids the uncomfortable question of what is outside the multiverse. Science fiction always characterizes the multiverse as finite, because the idea of an infinite multiverse is silly.

Another question Is what is between the universes of the multiverse. A common answer is that it's unknowable. But that's okay, because it's outside the universe, so we don't have to care.

What a messy view of reality! There can't be multiple everythings, there can only be one everything, otherwise it isn't everything. In my opinion, there can't be unknowable stuff outside the universe, because then you have everything plus other stuff, which again isn't everything. To have a "theory of everything" you need to have of concept of the universe whereby nothing is excluded.

Returning to science fiction, I know that a lot of contemporary movies and streaming TV shows rely on the concept of the multiverse, and its parallel universes. I watch and enjoy them too. The plot device of parallel dimensions (in one universe) would work equally well.

Criticizing the Cosmic Cocktail

1. It should be able to be summarized in a single paragraph.

2. It should not use vague expressions, which can't be formulated mathematically.

3. It should build on previous theories.

4. The "theory of everything" must combine general relativity and quantum theory.

5. The idea must be testable by an experiment.

6. The "simple underlying picture" must be understandable to a layman.

The cosmic cocktail recipe, I referred to earlier in chapter 14, is the established criteria for a theory of everything.[86] However, having solved the puzzle, it's obvious that an ingredient is missing. My cocktail tastes funny!

During a cable TV free preview weekend, I saw the series "Steven Hawking's Favorite Places" on Curiosity Stream. In one of the episodes, he discussed his thoughts on the boundary condition of the universe. At some point he proclaimed that the boundary condition is infinity. Then continued to say if the condition is infinity, then there is no boundary condition.[112]

This answer really bothers me. It doesn't feel right. The answer, it's just infinity is what scientists say in place of I don't know. What about a real hypothesis?

Primordial Stuff

What is the universe made of, or more properly from? Actualized quantum foam is a fun answer. For now, we'll just call it stuff. There three kinds of stuff:

- stuff of the observable universe
- stuff of the boundary condition
- stuff outside the universe

How can you have a theory of everything while there are three things instead of just one. There needs to be unity. One universe made from one stuff. The missing ingredient in the cosmic cocktail is the inclusion of a boundary condition that brings unity and answers to all remaining questions, like:

- What Is outside of the universe?
- Where Is the universe?
- What Is the universe made of?

The Boundary Condition

What if the boundary condition of the universe was collapsed spacetime via the Lorenz transformation? Essentially a zero-dimensional edge to our expanding universe. Being zero-dimensional, it would not need to be anywhere, as location would be meaningless. Thus, the idea of location would be essentially limited to the observable universe. It would make our universe an isolated system that can still be calculated finitely.

$$L = L_0 \sqrt{1 - \frac{v^2}{c^2}}$$

So how can you rationalize collapsed spacetime? Traditionally the Lorenz transformation is relativistically applied to only one axis of a three-dimensional Cartesian coordinate system. Specifically, the direction of travel. For example, if a spaceship was travelling near the speed-of-light along the x axis, it would be shortened by the Lorenz transformation, while leaving the y axis and the z axis untouched.[38] What if spacetime was relativistically moving in three-dimensions, such that all three axes will experience the effect of the Lorenz transformation. With enough energy, spacetime could collapse. I'm not sure if the right way to think of this is as a singularity shell or not. There certainly is enough untapped energy, in the form of zero-point vacuum energy, in the universe to account for this result. There may be no way to test this theory out directly, unless perhaps Hawking radiation can be detected coming from the cosmic event horizon.

Cosmic Calculator

If we return to the theory of everything, I presented in chapter 14, the universe was finitely calculated as a singular whole, in complete ignorance of the cosmic boundary condition. So, what now? This more sophisticated and complete totality is far more complicated. How should the total finite mass be calculated?

The universe has its own calculator, so there is no need for me to use mine. Since total applicable mass is determined empirically, no judgement is required as to how much "stuff" to include or exclude. Basically, the universe has done my homework for me! Problem solved.

The Tao

The benefit of this properly defined and complete theory of everything, is that now previously unanswerable questions can be answered.

Question: What Is outside of the universe?
Answer: Zero-dimensional spacetime, which extends over no distance.

Question: Where Is the universe?
Answer: Zero-dimensional spacetime doesn't require a location.

Question: What Is the universe made of?
Answer: It's composed of a single primordial stuff.

That last question might need to be unpacked more. The Asian philosopher Lao-tzu would be able to answer this one better. With a little paraphrasing and interpretation, he would say, that which is so primordial that it cannot be named, can only be referred to as the Tao.[113] As such, "the entity called the Tao existed before the universe came into being."[16]

I'm not really sure how I intellectually made the leap from chapter 14. It kind of just fell out of my head.

"Let your workings remain a mystery, just show people the results."[113]

- Tao-te Ching

Golden Paragraph Rewrite

Here is the theory of everything paragraph rewritten to incorporate the more complete final version:

The Standard Model of particle physics is almost a complete theory of everything, were it not for specific factors. These factors include that it fails to address dark energy, dark matter, and gravity as defined by general relativity. Extreme supermassive black hole magnetism can explain the behavior of galaxy movement over time in the universe, which is observed as dark energy, and supermassive black holes at the center of complex galaxies are also magnetically bonded to the constituent stars, resulting in the observation of dark matter. Resulting movement of the universe from this analysis results in the observation that the universe as a whole is acting like a harmonic oscillator. This movement is quantum mechanical. I propose that the universe is itself acting like a particle of physics with the properties of a graviton. The consequence of the universe being a graviton is unity between general relativity and the Standard Model within a single framework of quantum field theory. As such the speed-of-light value of "c" is defined relative to the total mass of the universe. This theory can be tested by using Electron Propulsion Engine technology to calculate the exact movement of the universe, by way of a gravity telescope. Additionally, Electron Propulsion can be used to measure the exact mass of the universe through superluminal performance data collected while the vehicle is in a naked wormhole. Therefore, when the "quantified" cosmic-scale graviton particle concept is added to the Standard Model, it's automatically a complete theory of everything. Additionally, the boundary condition is zero-dimensional spacetime resulting from a multidirectional Lorenz transformation contraction. Consequentially allowing for the calculative knowledge of what is both inside and "outside" of the universe, as a singular totality.

THE END

REFERENCES

1. Friedman, Michael et al. (2016). ChinaX Course Report. ChinaX Team. Harvard University.
2. Goulart, Justine. (2015). edX in 2015: Our Year in Review. EDX NEWS. edX.org. Retrieved from http://blog.edx.org/edx-year-in-review?track=blog.
3. Malan, David. (2016). CS50x Introduction to Computer Science. HarvardX. edXorg.
4. Corera, Gordon. (2012). WWII Pigeon Message Stumps GCHQ Decoders. BBC News. Retrieved from http://www.bbc.com/news/uk-20456782.
5. Grime, James. (2012). The Curious Case of the WWII Carrier Pigeon and the Unbreakable Code. Quite Easily Done. Cambridge University. iTunes U.
6. Meyer, David. (2014). NSA's quantum ambitions revealed in Snowden documents. CNN Money. Retrieved from https://gigaom.com/2014/01/03/nsas-quantum-ambitions-revealed-in-snowden-documents.
7. Morello, Andrea. (2013). How Does a Quantum Computer Work? Veritasium. University of New South Wales. Retrieved from https://youtu.be/g_IaVepNDT4.
8. Morello, Andrea. (2013). How to Make a Quantum Bit. Veritasium. University of New South Wales. Retrieved from https://www.youtube.com/watch?v=zNzzGgr2mhk.
9. Malan, David. (2013). This is CS50 2012: Shorts/RSA. Harvard University. iTunes U.
10. Vazirani, Umesh. (2013). CS-191x Quantum Mechanics and Quantum Computation. BerkeleyX. edX.org.
11. Vazirani, Umesh. (2011). A Computational Perspective on Quantum

Physics. UC Berkeley. Retrieved from https://youtu.be/jPZz5Bkh5lY.

12. Shumsker, Mark. (2012). How to Shuffle Cards. Truepokerdealer.com. Retrieved from http://truepokerdealer.com.

13. Bourdain, Anthony. (2013). Libya. Anthony Bourdain: Parts Unknown. CNN. Season 1, Ep. 6.

14. Melville, Herman. (1851). Moby Dick. Harper & Brothers. Ch. 38, 48, & 123.

15. Microsoft Encarta Encyclopedia. (2000)."Laozi." Microsoft Corporation.

16. Tzu, Lao. (1963). Tao Te Ching. Translated by D. C. Lau. Penguin Books. pp. 18, 101.

17. Liao, Waysun. (1995). The Essence of T'ai Chi. Shambhala. Boston & London.

18. Man-ch'ing, Cheng. (1981). T'ai Chi Ch'uan. North Atlantic Books, Berkeley. p.28.

19. Microsoft Encarta Encyclopedia. (2000)."Buddha." Microsoft Corporation.

20. Microsoft Encarta Encyclopedia. (2000)."Bodhidharma." Microsoft Corporation.

21. Microsoft Encarta Encyclopedia. (2000)."Zen." Microsoft Corporation.

22. Schiller, David. (1994). The Little Zen Companion. Workman Publishing. New York.

23. Weimin, Zhuang et al. (2013-14). 80000901_1X-80000901_2x History of Chinese Architecture: Part 1-2. TsinghuaX. edX.org.

24. The Holy Bible: King James Version. (1611). American Bible Society. New York.

25. Bol, Peter & Kirby, William. (2013-15). SW12.1x-SW12.10x China (Part 1-10). HarvardX. edX.org.

26. Meyer, Andrew. (2014). Cosmic Resonance Theory. ChinaX. HarvardX. edX.org.

27. Vergara, Alejandro. (2015). CEH.1-ENx Explaining European Paintings, 1400 to 1800. UC3Mx. edX.org.

28. Hawking, Stephen. (2001). The Universe in a Nutshell. Bantam Books. New York.

29. Martin, Walter & Ott, Magda. (2013). The Philosophy of Albert Einstein. Fall River Press. New York.

30. Hoffman, Jeffrey. (2015). 16.00x Introduction to Aerospace Engineering: Astronautics & Human Spaceflight. MITx. edX.org.

31. Dyson, George. (2002). Project Orion: The True Story of the Atomic Spaceship. Henry Holt and Company. New York.

32. Musk, Elon. (2017). Making Life Multiplanetary. SpaceX. International Astronautical Congress. Retrieved from https://youtu.be/tdUX3ypDVwI.

33. Mallove, Eugene & Matloff, Gregory. (1989). The Starflight Handbook: A Pioneer's Guide to Interstellar Travel. New York: John Wiley and Son, Inc.
34. Sagan, Carl. (1980). Cosmos. Random House. New York.
35. Conner, Paul H. (1996). Electron Propulsion Unit. U.S. Patent & Trademark Office. U.S. Patent No. 5,546,743. Retrieved from https://patentimages.storage.googleapis.com/04/61/f6/3c703e4069fafa/US5546743.pdf.
36. Conner, Paul H. (2017). Electron Propulsion Engine. U.S. Patent & Trademark Office. U.S. Patent No. 9,586,701. Retrieved from https://patentimages.storage.googleapis.com/0b/3b/bc/0e7fcfaa2f5b2e/US9586701.pdf.
37. Markusic, Thomas E. (2003-05). NASA, Marshall Space Flight Center, Propulsion Research Center.
38. Young, Hugh. (1992). University Physics, Eighth Edition: Extended with Modern Physics. Addison-Wesley Publishing Company, Inc.
39. Rassoul, Hamid. (2006). Florida Institute of Technology. Professor of Physics & Space Sciences. Director, Geospace Physics Laboratory. Associate Dean, College of Science.
40. Mangano, Michelangelo. (2005). CERN, PH-TH Department.
41. Strickland, Ashley. (2016). Mars mission astronauts could experience brain damage, study says. CNN. Retrieved from http://www.cnn.com/2016/10/13/health/mars-mission-astronaut-brain-damage/ index.html.
42. LaMotte, Sandee. (2021). Your body in space. CNN. Retrieved from https://www.cnn.com/interactive/2021/08/world/human-body-in-space-quiz-scn/.
43. Black, Phil & Smith-Spark, Laura. (2013). Russian meteor blast injures at least 1,000 people, authorities say. CNN. Retrieved from http://www.cnn.com/2013/02/15/world/europe/russia-meteor-shower.
44. Leopold, Todd et al. (2015). Halloween asteroid resembling skull narrowly misses Earth. CNN. Retrieved from http://www.cnn.com/2015/10/21/us/asteroid-earth-nasa-halloween-feat.
45. West, Andrew. (2014). ASTR105x Alien Worlds: The Science of Exoplanet Discovery & Characterization. BUx. edX.org.
46. Wall, Mike. (2013). It's Official! Voyager 1 Spacecraft Has Left Solar System. Space.com. Retrieved from http://www.space.com/22729-voyager-1-spacecraft-interstellar-space.html.
47. Rampino, Michael R. & Haggerty, Bruce M. (1996). The "Shiva Hypothesis": Impacts, Mass Extinctions, and the Galaxy. Springer Netherlands. Retrieved from http://link.springer.com/chapter/10.1007%2F978-94-009-0209-1_55.

48. Lafrance, Adrienne. (2015). The Chilling Regularity of Mass Extinctions. The Atlantic. Retrieved from http://www.theatlantic.com/science/archive/2015/11/the-next-mass-extinction/413884.

49. Mayell, Hillary. (2001). Chesapeake Bay Crater Offers Clues to Ancient Cataclysm. National Geographic News. Retrieved from http://news.nationalgeographic.com/news/2001/11/1113_chesapeakcrater.html.

50. Rajan, Nitya. (2015). Earth Could Be Hurtling Through Asteroid and Comet Shower That May Cause Mass Extinction Say Scientists. The Huffington Post UK. Retrieved from http://www.huffingtonpost.co.uk/2015/10/21/earth-could-be-hurtling-through-asteroid-and-comet-shower-that-may-cause-mass-extinction-say-scientists-_n_8344510.html.

51. Nilsen, Ella & Marsh, René. (2022). US scientists reach long-awaited nuclear fusion breakthrough, source says. CNN. Retrieved from https://www.cnn.com/2022/12/12/politics/nuclear-fusion-energy-us-scientists-climate.

52. Strickland, Ashley. (2022). The DART mission successfully changed the motion of an asteroid. CNN. Retrieved from https://www.cnn.com/2022/10/11/world/nasa-dart-success-update-scn.

53. Achenbach, Joel. (2017). How Mars lost its atmosphere, and why Earth didn't. The Washington Post. Retrieved from https://www.washingtonpost.com/news/speaking-of-science/wp/2017/03/30/how-mars-lost-its-atmosphere-and-why-earth-didnt/?noredirect=on&utm_term=.3e929b20f8a3.

54. Factor, Sam. (2015). Is there a way to provide a magnetic field for Mars? Ask an Astronomer. Retrieved from http://askanastronomer.org/planets/2015/11/20/can-we-create-a-magnetic-field-for-mars/.

55. Motojima, Osamu & Yanagi, Nagato. (2008). Feasibility of Artificial Geomagnetic Field Generation by a Superconducting Ring Network. National Institute for Fusion Science (NIFS) of Japan. Retrieved from http://www.nifs.ac.jp/report/NIFS-886.pdf.

56. Wehner, Mike. (2018). Earth's days used to be just 18 hours long, but the Moon changed that. BGR. Retrieved from https://bgr.com/2018/06/06/earth-moon-days-length-history/.

57. Cain, Fraser. (2008). Mars Tilt. Universe Today. Retrieved from https://www.universetoday.com/14894/mars-tilt/.

58. Strickland, Ashley. (2020). Mars' moons may hint that the planet once had rings. CNN. https://www.cnn.com/2020/06/03/world/mars-rings-moons-scn/index.html.

59. NASA Exoplanet Archive. (2016). NASA Exoplanet Science Institute.

California Institute of Technology. Retrieved from
http://exoplanetarchive.ipac.caltech.edu.

60. Byrd, Deborah & Imster, Eleanor. (2019). Today in Science: 1st Planet
    Orbiting A Sun-Like Star. EarthSky. Retrieved from
    https://earthsky.org/space/this-date-in-science-first-planet-discovered-
    around-sunlike-star.

61. Newsome, John. (2015). Space anomaly gets extraterrestrial intelligence
    experts' attention. CNN. Retrieved from http://www.cnn.com/2015/
    10/15/world/extraterrestrial-intelligence-anomaly/index.html.

62. King, Bob. (2015). What's Orbiting KIC 8462852 – Shattered Comet or
    Alien Megastructure? Universe Today. Retrieved from
    http://www.universetoday.com/122865/whats-orbiting-kic-8462852-
    shattered-comet-or-alien megastructure.

63. Newsome, John. (2015). NASA says space anomaly likely caused by
    comets. CNN. Retrieved from http://www.cnn.com/2015/11/25/
    us/nasa-space-anomaly-comets.

64. West, Andrew. (2014). ASTR105x Alien Worlds: Office Hours #4.
    BUx. Retrieved from https://www.youtube.com/watch?
    v=XrH_oSfFJRA.

65. Drake, Frank. (2010). A Life with SETI. Astrobiology and Space
    Exploration. Stanford University. iTunes U. Ep.16.

66. Mack, Eric. (2015). Could a 'super-Earth' be even more habitable than
    our own planet? CNET. Retrieved from http://www.cnet.com/
    news/a-super-earth-could-be-even-more-habitable-than-our-planet.

67. Sasselov, Dimitar. (2015). SPU30x Super-Earths & Life. HarvardX.
    edX.org.

68. West, Andrew. (2014). ASTR105x Alien Worlds: Office Hours #3.
    BUx. Retrieved from https://www.youtube.com/watch?
    v=fnRQ5tSQZ0c.

69. Schmidt, Brian & Francis, Paul. (2014). ANU-ASTRO1x Greatest
    Unsolved Mysteries of the Universe. ANUx. edX.org.

70. Heilpern, Will. (2015). Wolf 1061 exoplanet: 'Super-Earth' discovered
    only 14 light-years away. CNN. Retrieved from http://www.cnn.com/
    2015/12/17/world/wolf-1061-exoplanet-alien-life.

71. Press Release. (2015). Potentially habitable super-Earth found just 14
    light-years away. University of New South Wales. Astronomy Now.
    Retrieved from https://astronomynow.com/2015/12/18/
    potentially-habitable-super-earth-found-just-14-light-years-away.

72. Strickland, Ashley. (2016). Closest potentially habitable planet to our
    solar system found. CNN. Retrieved from
    http://www.cnn.com/2016/08/24/health/proxima-b-centauri-rocky-
    planet-habitable-zone-neighbor-star/index.html.

73. Strickland, Ashley. (2020). Astronomers confirm Earth-size exoplanet

around nearest star and maybe more. CNN. Retrieved from https://www.cnn.com/2020/06/04/world/proxima-b-proxima-c-exoplanets-scn/index.html.

74. Bailyn, Charles. (2007). Astrophysics: Frontiers and Controversies. Yale University. iTunes U. Ep. 8-24.

75. Hawking, Stephen. (1993). Black Holes and Baby Universes and Other Essays. New York: Bantam Books.

76. Lewin, Walter. (2008). Physics I: Classical Mechanics. MIT. iTunes U. Ep. 19.

77. Kaku, Michio. (2008). Physics of the Impossible. New York: Doubleday.

78. Lucas, George. (1977). Star Wars: Episode IV: A New Hope. 20th Century Fox.

79. Hurley, Dan. (2020). The Quantum Internet will blow your mind. Here's what it will look like. Discover Magazine. Retrieved from https://www.discovermagazine.com/technology/the-quantum-internet-will-blow-your-mind-heres-what-it-will-look-like?utm _campaign=Feed%3A +AllDiscovermagazinecomContent+%28All +DISCOVERmagazine.com+stories%29&utm_medium=feed&utm_s ource=feedburner.

80. Emspak, Jesse. (2016). Quantum Entanglement: Love on a Subatomic Scale. Space.com. Retrieved from http://www.space.com/31933-quantum-entanglement-action-at-a-distance.html.

81. Microsoft Encarta Encyclopedia. (2000). "Eratosthenes." Microsoft Corporation.

82. O'Connor, J. J. & Robertson, E. F. (1999). Eratosthenes of Cyrene. University of St Andrews, Scotland. Retrieved from http://www-history.mcs.st- andrews.ac.uk/history/Mathematicians/ Eratosthenes.html.

83. Donovan, Dennis P. (1996). Calculating the Circumference of the Earth. Rice University. Retrieved from http://www.math.rice.edu/ ~ddonovan.

84. Burchat, Patricia. (2008). Patricia Burchat Sheds Light on Dark Matter. Physics: The Edge of Knowledge. iTunes U.

85. Puthoff, Harold. (1990). Everything for Nothing. New Scientist. Retrieved from https://www.newscientist.com/article/mg12717275-500.

86. Kaku, Michio. (2005). What to Do If You Have a Proposal for the Unified Field Theory? MKaku.org. Retrieved from http://mkaku.org/home/articles/what-to-do-if-you-have-a-proposal-for-the-unified-field-theory.

87. Zeidler, Sari. (2012). The End of the Galaxy as We Know It? CNN. lightyears.blogs.cnn.com. Retrieved from

http://lightyears.blogs.cnn.com/2012/05/31/the-end-of-the-galaxy-as-we-know-it.

88. Binney, James. (2009). Quantum Mechanics. Oxford University. iTunes U. Ep. 007.
89. CERN. (2008). European Organization for Nuclear Research. CERN. Retrieved from http://public.web.cern.ch/public.
90. Cox, Brian. (2008). Brian Cox on CERN's Supercollider. Physics: The Edge of Knowledge. iTunes U.
91. CERN: Science. (2008). The Large Hadron Collider in 10 Minutes. CERN. iTunes U.
92. Greene, Brian. (2005). Brian Greene String Theory. Physics: The Edge of Knowledge. iTunes U.
93. Isaacson, Walter. (2007). Einstein: His Life and Universe. New York: Simon & Schuster.
94. Gell-Mann, Murray. (2007). "Murray Gell-Mann on Beauty and Truth in Physics." Physics: The edge of Knowledge. iTunes U.
95. Lewin, Walter. (2008). Physics II: Electricity & Magnetism. MIT. iTunes U. Ep. 25.
96. Charles, R.H. (2003). The Book of Enoch the Prophet. San Francisco: Weiser Books.
97. Reps, Paul & Senzaki, Nyogen. (1994). Zen Flesh, Zen Bones. Boston: Shambhala.
98. Kaviratna, Harischandra. (1997). The Heart Sutra: Prajnaparamita-Hridaya-Sutra. Theosophical University Press. Retrieved from http://www.theosophy-nw.org/theosnw/world/asia/as-heart.htm.
99. Robinson, James. (1978). The Nag Hammadi Library. Harper: San Francisco.
100. Francis, Paul. (2014). Professor of Astronomy & Astrophysics. Australian National University.
101. ScienceAtNASA. (2014). ScienceCasts: The Milky Way is Not Just a Refrigerator Magnet. NASA. Retrieved from https://youtu.be/HxqctkvlTtw.
102. Byrd, Deborah. (2019). Our Milky Way is Warped. EarthSky. Retrieved from https://earthsky.org/ space/milky-way-warped-twisted-study-cepheids.
103. Dvorsky, George. (2019). Another Study Finds Our Galaxy Is 'Warped and Twisted.' Gizmodo. Retrieved from https://gizmodo.com/another-study-finds-our-galaxy-is-warped-and-twisted-1836881811.
104. Starr, Michelle. (2019). The Latest Calculation of Milky Way's Mass Just Changed What We Know About Our Galaxy. ScienceAlert. Retrieved from https://www.sciencealert.com/the-most-accurate-measurement-yet-of-the-milky-way-s-mass-puts-us-ahead-of-andromeda.
105. Siegel, Ethan. (2019). Could the Milky Way Be More Massive Than

Andromeda? Medium. Retrieved from https://medium.com/starts-with-a-bang/could-the-milky-way-be-more-massive-than-andromeda-38725096d016.

106. Mao, Sui Ann. (2017). Distant galaxy sheds light on how magnetism formed in the early Universe. Max Planck Institute for Radio Astronomy. Retrieved from https://www.mpg.de/11454790/magnetic-fields-form-early-in-life-of-galaxy.

107. Temming, Maria. (2019). In A First, Magnetic Fields Have Been Spotted Between Two Galaxy Clusters. Science News. Retrieved from https://www.sciencenews.org/article/magnetic-fields-between-galaxy-clusters.

108. Kierdorf, Maja. (2017). Giant Magnetic Fields in The Universe. Max Planck Institute for Radio Astronomy. Retrieved from https://www.mpifr-bonn.mpg.de/pressreleases/2017/4.

109. Buhrke, Thomas. (2017). Forces That Rule Galaxies. Max Planck Institute for Radio Astronomy. Retrieve from https://www.mpg.de/9093617/F003_Focus_034-041.pdf.

110. Carpineti, Alfredo. (2019). Supermassive Black Holes May Have Formed Without Any Stars. IFLScience. Retrieved from https://www.iflscience.com/space/supermassive-black-holes-may-have-formed-without-any-stars/.

111. Foster, Brendan Z. (2019). Are We All Wrong About Black Holes? Wired. Retrieved from https://www.wired.com/story/are-we-all-wrong-about-black-holes/.

112. Hawking, Stephen. (2016). Stephen Hawking's Favorite Places. Curiosity Stream. Ep. 3.

113. Tzu, Lao. (2007). Tao Te Ching. Translated by Stephen Mitchell. HarperAudio. Ch. 1 & 36.

APPENDIX

U.S. Patent 9,586,701

Conner Creations, LLC Website